Enterprise GIS
Concepts and Applications

Enterprise GIS
Concepts and Applications

by
John R. Woodard MS, GISP

CRC Press
Taylor & Francis Group
Boca Raton London New York

CRC Press is an imprint of the
Taylor & Francis Group, an **informa** business

CRC Press
Taylor & Francis Group
6000 Broken Sound Parkway NW, Suite 300
Boca Raton, FL 33487-2742

First issued in paperback 2022

© 2020 by Taylor & Francis Group, LLC
CRC Press is an imprint of Taylor & Francis Group, an Informa business

No claim to original U.S. Government works

ISBN 13: 978-1-03-247494-6 (pbk)
ISBN 13: 978-1-138-47829-9 (hbk)
ISBN 13: 978-1-351-06900-7 (ebk)

DOI: 10.1201/9781351069007

This book contains information obtained from authentic and highly regarded sources. Reasonable efforts have been made to publish reliable data and information, but the author and publisher cannot assume responsibility for the validity of all materials or the consequences of their use. The authors and publishers have attempted to trace the copyright holders of all material reproduced in this publication and apologize to copyright holders if permission to publish in this form has not been obtained. If any copyright material has not been acknowledged, please write and let us know so we may rectify in any future reprint.

Except as permitted under U.S. Copyright Law, no part of this book may be reprinted, reproduced, transmitted, or utilized in any form by any electronic, mechanical, or other means, now known or hereafter invented, including photocopying, microfilming, and recording, or in any information storage or retrieval system, without written permission from the publishers.

For permission to photocopy or use material electronically from this work, please access www.copyright.com (http://www.copyright.com/) or contact the Copyright Clearance Center, Inc. (CCC), 222 Rosewood Drive, Danvers, MA 01923, 978-750-8400. CCC is a not-for-profit organization that provides licenses and registration for a variety of users. For organizations that have been granted a photocopy license by the CCC, a separate system of payment has been arranged.

Trademark Notice: Product or corporate names may be trademarks or registered trademarks, and are used only for identification and explanation without intent to infringe.

Publisher's Note
The publisher has gone to great lengths to ensure the quality of this reprint but points out that some imperfections in the original copies may be apparent.

Visit the Taylor & Francis Web site at
http://www.taylorandfrancis.com

and the CRC Press Web site at
http://www.crcpress.com

Dedication

This book is dedicated to my wife and my parents.

For their endless love, support, patience, and

encouragement during this writing process.

Contents

List of Figures xi
List of Tables xiii
Preface xv
Acknowledgments xvii
Author xix

1. **Introduction and Organization** 1
   1.1 Introduction 1
   1.2 Information sources 6
   1.3 Enterprise architecture 9
   1.4 Conclusion 12
       1.4.1 Organization 12
   1.5 Applied GIS 13
   1.6 Assignment 14
       1.6.1 Assignment 1 Article Review 14
   Bibliography 14

2. **Enterprise Geographic Information System (Enterprise GIS)** 17
   2.1 Just what is enterprise GIS? 18
   2.2 Why are there so many different, yet similar, definitions for enterprise GIS? 18
   2.3 Case 1: Medina County Health Department, Medina County, Ohio 19
   2.4 A word on professionalism 21
   2.5 The enterprise GIS 25
   2.6 The beginning of the enterprise system 26
   2.7 Summary 29
   2.8 Assignment 29
   Bibliography 30

3. **Enterprise Architecture** 31
   3.1 What is enterprise architecture? 31
   3.2 Capabilities 33
   3.3 Flexible architecture 35
   3.4 Designing an enterprise architecture 37
   3.5 Establishing goals 38
   3.6 Case 2: The City of Oberlin, County of Lorain, Ohio 41
   3.7 Summary 42

	3.8	Assignment	42
		Assignment 3: The company	42
		Assignment 3.1 Establishing goals	42
	Bibliography		42

4. Roles in Enterprise Architecture 45
  4.1 Roles and responsibilities 45
  4.2 Building the team 50
  4.3 GIS manager/enterprise architect 51
  4.4 Business architect 52
  4.5 Data architect 53
  4.6 Data steward 53
  4.7 Subject matter expert 54
  4.8 Applications/technology architect 55
  4.9 Owner and executive management 56
  4.10 Department supervisors 57
  4.11 Summary 57
  4.12 Assignment 58
         Assignment 4: Why 58
         Assignment 4.1: Roles 58
  Bibliography 58

5. Data Architecture 61
  5.1 Data architecture 61
  5.2 Data vs. information 62
  5.3 Enterprise data 62
  5.4 Metadata 64
  5.5 System hierarchies 64
  5.6 Data standards 66
  5.7 Reference data 72
  5.8 Data formatting 73
  5.9 Geodatabase domains 74
  5.10 Summary 75
  5.11 Assignment 75
         Assignment 5: Data examples 75
  Bibliography 75

6. The EA/Enterprise GIS Toolbox 77
  6.1 EA/EGIS tools 77
         6.1.1 Business on a page (BOAP) 78
         6.1.2 Workflow model 81
         6.1.3 UML activity diagram 81
         6.1.4 Strategic analysis methods 82
         6.1.5 Backcasting 83

	Case of the City of Strongsville	84
	City of Strongsville enterprise GIS build timeline	87
6.2	Present/future state model	87
6.3	Process model	88
6.4	Improvising tools	88
6.5	Final comment	89
6.6	Summary	89
	Bibliography	89

7. Designing an Enterprise GIS with EA Tools 91

7.1	Designing the enterprise GIS with EA tools	91
7.2	Enterprise GIS goals	91
7.3	The geodatabase	92
7.4	Catalogs	96
7.5	Data catalogs	96
7.6	Technology catalog	98
7.7	Application catalog	101
7.8	Current state to future state model	104
7.9	Summary	110
7.10	Assignment	111
	Assignment 7.1: Build a data, technology, applications catalog	111
	Assignment 7.2: Catalog analysis	111
	Assignment 7.3: Building a model for current state to future state	111
	Assignment 7.4: Build a maturity model	111
	Bibliography	112

8. System Visualization 115

8.1	Strategic architecture	115
8.2	Sub-architecture	119
8.3	Summary	123
8.4	Assignment	123
	Assignment 8.1: Visualizing the entire system	123
	Assignment 8.2: Visualization of current state	123
	Assignment 8.3: Visualization of sub-architecture	123
	Assignment 8.4: Visualization of future state	124
	Bibliography	124

9. Governance 125

9.1	What is governance?	125
	Case: The City of Oberlin, Lorain County, Ohio	126
	Case: The City of Oberlin, Lorain County, Ohio	127
	Case: Medina County Health Department (MCHD), County of Medina, Ohio	128

	9.2	Governance policy 129
		Case: City of Strongsville, County of Medina, Ohio 130
	9.3	Governance review board 130
		Case: The City of Oberlin, Lorain County, Ohio 131
	9.4	Compliance waivers 132
		List of requirements for compliance 133
		Extent 133
		Items required for compliance 133
		Actions needed 133
		What are the consequences? 133
		Sample response to the compliance waiver 133
		Actions needed 134
		What are the consequences? 134
		Penalties 134
		Case: The City of Avon, Lorain County, Ohio 134
	9.5	Summary 135
	9.6	Assignment 136
		Assignment 9: Define governance 136
	Bibliography 136

10. Conclusion and Future of GIS 137
    10.1 Conclusion 137
    10.2 Leadership 140
    10.3 Future directions 141
    10.4 Final thoughts 143
    Bibliography 143

Appendix 145

Index 149

List of Figures

Figure 3.1	Example of a goals chart	39
Figure 4.1	Roles and responsibilities	47
Figure 6.1	Example of a BOAP	79
Figure 7.1	Example of enterprise GIS goals chart	93
Figure 7.2	Example of maturity table	94
Figure 7.3	Example of data catalog	97
Figure 7.4	Example of tech catalog	100
Figure 7.5	Example of applications catalog	103
Figure 7.6	Current state to future model	106
Figure 7.7	Maturity growth chart	108
Figure 8.1	OTC overview architecture	117
Figure 8.2	Geodatabase architecture model	120
Figure 8.3	Data connections model	121

List of Tables

Table 4.1 GIS Department Roles and Responsibilities 48

Table 4.2 Combined GIS and EA Roles and Responsibilities 49

Table 4.3 City of Oberlin Enterprise GIS Organizational GIS Roles 50

Table 5.1 Hierarchy of WoodWeyr, LLC 65

Table 5.2 Reference Card Template 68

Table 5.3 Lyndhurst Street Centerline Reference Card 69

Table 5.4 Lyndhurst Flood Damage Reference Card 70

Table 5.5 Geodatabase Domain for Pipe Material 74

Table 7.1 Geodatabase Domain for Pipe Material 95

Preface

Many people have asked, "Why?" Why this book about enterprise architecture for geographic information systems? ESRI has already created these systems! There are many books about enterprise architecture and operations.

These are valid questions, and they deserve answers.

This textbook is intended as a helpful aid for GIS professionals and users searching for ways to improve themselves and information systems. This book is a way of giving back to the people who taught, encouraged, and sometimes provided a proper kick to "you know where," when this author required proper motivation.

These same people (professors, teachers, Seabees, farmers, and laborers) also instilled the professional ideals of hard work, responsibility, accountability, helping others, and pride, in an inexperienced newbie. This book also seeks to honor their efforts, with a tool for tomorrow's professionals.

Smoothing the path for those behind you is one of those professional responsibilities everyone seems to discuss, but only a few people practice. This book will hopefully help smooth the rough spots in the paths of future enterprise GIS builds and ensure a smoother transition from a simple desktop to an organizational information system. The author's understanding and methodologies in developing enterprise systems come from developing several municipal GIS systems and a second master's degree in digital science concentrated upon enterprise architecture.

Continuing education is one of the secrets to becoming a well-rounded professional. Scuba divers have a magazine *Dive Training*, which has the motto "A good diver is always learning!" This author has not only lived by, and trained for scuba following, the slogan but adapted it for the field of GIS, "A good GISer, is always learning!" This phrase can be applied or act as a slogan for any professional regardless of profession.

Acknowledgments

First and foremost, I would like to thank my wife for her love and willingness to sacrifice "our time" to enable me to finish writing this project.

I would like to thank my parents for their support and help; they were there when I needed them.

A very special thank you to Dr. Tyner! Without his suggestions and encouragement this book would never have been written.

Thank you to the partners/owners of my former company, Chagrin Valley Engineering, Ltd. (CVE). They provided support, encouragement, and the space to fail, when developing enterprise GIS. They graciously allowed me to use CVE project sheets as appendices for this book.

I would like to say thank you to my current employer Davey Resource Group, Inc. (DRG) and The City of Oberlin, Ohio Public Works Department for their support and for allowing me to use Oberlin's project sheet as an appendix for this book.

A special thank you to Ms. Tabitha Messmore for being a "go-to" mentor and advisor.

A special thank you to Mrs. Debbie Sheeler for her encouragement and support.

A special thank you to Dr. Denise Bedford who introduced me to the field of enterprise architecture and who provided timely insights, encouragement, and templates for many enterprise and business architecture models.

Thank you to all my friends, who encouraged me along the way and never let me forget how far I had come, during moments of self-doubt. Your friendship has provided me with many wonderful adventures, experiences, and memories that helped me through the hard times.

Thank you, Lord, for providing me with all these wonderful people in my life, and for always being there for me.

John R. Woodard MS, GISP

Author

John R. Woodard, MS, GISP has worked in the field of geographic information systems (GIS) since 2002. He was fortunate enough to be offered a position with the Summit County Health Department's Community Health Division. John was responsible for planning, implementing, and coordinating all GIS activities for the Health Department. A high point in his time with Summit County Health was managing a multi-agency project composed of nurses, administrators, and first responders to create an interactive disaster/terrorism response map. This interactive map was designed to provide first responders information about nearby facilities and assets in a time of emergency.

He spent 10 years with Chagrin Valley Engineering, forming a close working relationship with clients, engineers, and field crews. This is where he became interested in working with enterprise architecture as a foundation for enterprise GIS systems. He was responsible for designing, building, and maintaining enterprise GIS systems for 14 municipalities in Northeast Ohio. His excellent work for The City of Avon was recognized with a GIS Best Practice award from the Ohio Geographically Referenced Information Program (OGRIP) in 2016.

Mr. Woodard firmly believes in continuing education and helping others. Since May 2003, he has held the position of Adjunct Faculty at the Northeast Ohio Medical University, Rootstown, Ohio. During this time, he completed a master's degree in GIS/Geography from the University of Akron and a master's degree in digital science concentrated in enterprise architecture from Kent State University, Kent, Ohio. He was awarded his certification as a GIS Professional (GISP) in 2015.

John has had a varied career outside of the field of GIS. He serviced with the United States Navy Reserve Seabees from 1995 to 2005 and has pursued the hobbies of scuba diving and beekeeping. He currently is employed with Davey Resource Group Ltd as a Project Developer for the GIS/IT division. John lives at home in Ohio with a beautiful wife and 20-month-old daughter.

1

Introduction and Organization

1.1 Introduction

This book does not advocate "scrapping" enterprise geographic information system (EGIS) systems but will demonstrate how GIS benefits through the use of enterprise architecture (EA) theories, ideas, and methodologies. Many people feel that when more than one computer is connected to a dataset, sharing files and printers between people or departments constitutes an enterprise system. This is far from the truth, particularly in the field of GIS, where people play fast and loose with file-naming conventions, organization, or formatting. These systems are plagued with improper file names, inefficient or duplicated pathways, incorrect data, or damaged files, and there is the potential for many other problems.

Any information system that has these problems will not succeed. The system will be inefficient, slow to respond, and provide bad data when making decisions. Problems like duplicated pathways to files can dramatically slow system response time, yet remain hidden from a low-level scan. Over time, these problems will have a cumulative impact on system performance, ultimately leading to failure. Attempting to work with a system containing flaws would be the same as shooting oneself in the foot.

People only seem to notice a system's problems after it has been overwhelmed by preventable ones. Naturally, system failures always seem to happen during a "crisis or rush" situation where time is of the essence! A system that fails will have a decidedly negative impact on a user's attempt to meet deadlines. Failure to meet a critical deadline will have a negative impact not only on the system, but on the professional depending on the system! Yes, everyone will miss a deadline once in a while and system failure will be accepted as a justifiable reason, once, or maybe twice. Repeated failures to meet deadlines will always fall back on the system user and blaming the system at that time will be viewed as nothing more than an excuse for their incompetence.

Information system failures will at the very least damage a person's professional reputation, lead to being fired, or both.

System problems can be prevented, and reputations and jobs saved with the use of ideas and methods from the field of enterprise architecture! The best solution for solving problems is to prevent them from reaching the system. Enterprise architecture promotes a strategic plan for eliminating and preventing problems from occurring in an organization's business systems. The purpose of enterprise architecture is to create a strategic plan that brings all of an organization's assets, staff, and resources into one effort that supports the primary business mission.

A GIS is a powerful tool that supports the primary business mission. Poorly designed or implemented GISs are always plagued with problems in data, communication, duplicated efforts, files, processes, etc. These problems will cripple the functionality of any desktop or enterprise GIS. The author would like the reader to consider this truth that behind every successful information system is a well-executed enterprise architecture (plan). Enterprise architecture holds the keys to creating successful GISs. A few keys are strategic architecture, data architecture, data formats, standards, etc. The strategic or enterprise architecture is the keystone that begins and ties this system foundation together. Building a strong system foundation will require a thorough understanding of the similarities and differences between enterprise GIS and enterprise architecture.

A GIS professional can use these design keys to build enterprise GISs that are improved or enhanced with ideas from enterprise architecture. Enterprise GISs that are enhanced with enterprise architecture elements save money, provide a high return on investment, and operate with a high level of efficiency. GIS professionals can create information systems that are free of system errors, dependable, and filled with trusted information. The person who is responsible for developing and deploying a system that eliminates the problems of the "old way," at a lower cost with a higher return on investment (ROI), is guaranteed to look like a superstar!

Enterprise GISs are primarily designed and implemented to provide GIS capabilities to departments and members of an organization. Unfortunately, too often, the users and managers of a GIS fail to provide an overarching goal or strategic purpose. Commercial software vendors, conference presentations, and academic authors' papers and publications champion the virtues of an enterprise system. Many organizations simply lack the funding and staff resources to design and implement an enterprise system. Commercial software companies realize this fact, and promote the idea that their software can create an "out of the box" solution with a simple mouse click. These software companies peddling these "easy and simple" solutions have the goal of making a sale and locking the organization into a proprietary system. Proprietary software means that any updates, changes, or new applications must work (for a fee) with one company's application. The client is forced to maintain a business relationship with this vendor regardless of the level of satisfaction with the product or service. Locking the client into a proprietary solution means a steady stream of revenue for the software company.

Academic presentations, publications, or papers always seem to deal with an idealistic world. It is a world where everyone has the finances, time, and manpower to simply create an organized system out of thin air! Academic people tend to believe that everyone has or soon will deploy an enterprise system. These papers and presentations promote the impression that everyone has developed a system or that it is fairly easy and quick to design the system. Everyone, commercial software vendors, consultants, or academic professionals overlook discussing and explain the process to develop and distribute a GIS with very little "how-to" details of system design.

There are GISs that are simply distributed throughout an organization. These are created when different departments or several people start working with GIS. This is not a true enterprise system or even a good way of sharing files. The problem with this "distributed" GIS is that if each person or department produces maps and data with no sense of organization or mutual project support, this ad hoc distributed GIS results in duplication of efforts, data, and results. There is chaos within the system as people produce large amounts of information without a sense of direction or value. Information that has no value is useless, which in turn means that all of the effort, time, and financial resources expended to produce this information was wasted.

Consider a county auditor's dataset with information on land parcels within the county. Every week the parcel information (lines, ownership, etc.) must be updated. A data architecture to standardized or maintain data quality was never implemented. This lack of control has resulted in a dataset filled with problems. A lack of standardization of the names of parcel owners is one example of a data problem. A search of the parcel dataset revealed that municipal names were misspelled, incorrectly abbreviated, even listed as a village, or township instead of a city. Incorrect names for ownership create a situation where a query against the database will result in the wrong information being obtained. The wrong information, in turn, guarantees that any analysis performed on the corrupted information will lead to a bad result. A repeated pattern of poor decisions caused by bad information will result in a system that is no longer trusted.

Property boundaries (or parcel lines) populate the system with a very unique problem. The County Engineers' office has the responsibility of ensuring that all of the boundaries defining property parcels are correct. However, the county auditor is tasked with maintaining all tax records, ownership, and related information to the property parcels. Two departments maintaining different types of information for the same dataset will have to deal with, if not a problem, miscommunication.

When someone decides to combine or split a property parcel, the auditor's staff updates the increase or decrease in the tax value of the parcel and then uses their GIS to modify the property. A simple change in the Auditor's GIS will either split the parcel into two pieces or combine two parcels into one new parcel. This GIS process takes only a few minutes to execute, everything is updated, new tax values, and ownership is updated! Amazing, it is all finished quickly ... or is it? The auditor's parcel dataset has been updated with new information and even new boundaries for the properties impacted by the purchase. The reader should recall the fact that the engineer is responsible for maintaining the property boundaries!

This is where the problem occurs between the two departments. Legally, the engineer's office cannot accept or make changes to any property boundary without a proper land survey. A legal process exists in every county that must be followed by the engineer's office to ensure that property lines are properly measured, recorded, and certified by a professional surveyor. This process prevents the engineer's staff from simply deleting or adding lines to a parcel map.

This creates the problem of two maps within the county depicting property information. One map represents the county as the auditor's, who views properties via the tax information. The other map represents the county as the engineer's, who views the properties via a certified survey. The auditor's tax map will have all of the updated information for taxes, ownership, etc., but the parcel lines will not be legally correct. The engineer's map naturally has the opposite situation, the parcel lines are correct, but all of the information for taxes, ownership, etc., will be out of date. This situation creates a huge amount of confusion for county planners, private developers, and members of the public.

An answer to the two-map situation is a data plan (architecture) that addresses the "how to" for updating parcel information. The best solution will be a process that allows both departments to edit the GIS parcel map simultaneously in real time. The process would require checks and balances. The first system check consists of a notice to the engineer that a change in a property boundary is requested. This request would begin a review of the property boundaries that will determine the amount of change required. A list of recommended or approved surveyors would be sent to the citizen initiating the process. The auditor's staff could begin the update for the citizen with the understanding that nothing is approved or legally changed until the survey has been filed, approved, and accepted by the county engineer. This solution will result in accurate information for the tax map, while improving communication and coordination between the auditor's and engineer's departments.

Enterprise architecture focuses on providing organization and managing risk and change within a complex system. An organization's architecture only succeeds if it achieves the total support or buy-in from departments and members (Wilson 2013). Clear and concise communication is critical to achieving buy-in from company stakeholders, and will enable the system to manage risk and adapt to today's rapidly changing digital world.

Company stakeholders, managers, and staff will not merely champion a new system on a designer's word. Anyone who wishes to create and implement an enterprise system, especially an EGIS, must build trust with everyone who will interact with the system. One can describe how the proposed system will save everyone time, respond quickly to queries, or support business analysis, etc. A financial analysis justifying the system through cost savings in labor, time, and resources, resulting in rising profits will result in the enterprise system getting a second look. Clear and concise communication can quickly explain to everyone how they all benefit by supporting a new system.

Everyone seems to understand the numbers of cost and time. Cost and time are intertwined and form two of the biggest issues when deploying a new system. However, they are your allies in defending the need for a system overhaul or build. This method of defending the need or cost of developing a new system is illustrated through the author's own experience. Your author was asked to provide "real" numbers that a health department's board would easily understand. The author's approach was to break down the health department's total expense for their personnel to inspect a septic system. First, the person inspecting was asked how many septic systems he inspected per day or week. The answer of four per week was very surprising!

The health department's staff explained that searching through the old file cabinets (no digital records) took a lot of time. Two administrative assistants estimated that they each spent four hours researching the septic systems that needed inspection. The inspector estimated that he spent roughly four to six hours, reviewing their research, using Google Maps to view, and then determine the best route to the property parcel with the septic system. This meant that each septic system to be inspected required 8 plus 6 for a total of 14 man hours to simply research and locate the system. Therefore, the department was investing 56 hours (4 × 14) to inspect four septic systems! Let's assume that the average hourly rate for each person on the project is $20.00 per hour. The total cost for four septic systems would be $1,120.00 and that is before the inspector leaves the office to inspect the septic system.

The author made the calculation that once an enterprise GIS was properly created and installed, the health department's costs would show a dramatic drop. Administrative assistants would no longer need to sort through old file cabinets and handwritten notes. A simple click on the parcel would display hyperlink documentation for the parcel and septic system (time invested five minutes). They would then be able to review the information, print, or email the findings to the inspector. The inspector would no longer have to read through a book of documents at his desk, a simple mouse click would show all the information that required review (20 minutes per review). Now he could switch to the enterprise GIS link to Google Maps and simply click on route to determine the quickest path to the septic system to be inspected (30 minutes total). Now the total manhours expended is: 5 minutes (×2) = 10 + 20 + 30 = 60 minutes or 1 hour per inspection. Now the health department invested cost would be $20.00 × 1 hour = $20.00 per inspection setup.

When the client can see their costs drop from $1,120.00 to $20.00 per inspection many of the arguments against the new system will disappear.

Agility, flexibility, adaptability, and finding new ways to beat the competition to the profit is every businessman's goal. One of the benefits of enterprise architecture is more clear lines of communication throughout the organization. A hierarchy of communication that defines who contacts who will eliminate the needless directory searches that cost time, money, and slow the company's reaction time to market changes. A communication hierarchy enables decision-makers to learn about business problems or opportunities, assess the situation, make decisions, and then direct the company's response to the circumstances.

There is a belief that if more than one department has access to GIS capabilities, then the company has developed an EGIS. This claim is accurate if one accepts the idea that the EGIS only provides tools to all levels of an organization. Unfortunately, this type of EGIS lacks the planning foundation that would give the system organization and direction. This system will contain many errors, duplicated data, redundant projects, etc. It will not be efficient and trusted, nor will it provide reliable support for business decisions.

The discipline of enterprise architecture can provide the organizational features that are missing from an enterprise GIS. Adding the theories, methodologies, and visions of enterprise architecture to enterprise GIS will align the system to support the organization's goals. A skillful merger of enterprise architecture and GIS will bring order to a system in chaos.

The purpose of this book is to bridge the divide between the practitioners of enterprise architecture and GISs. Readers will learn some of the concepts, ideas, and methodologies of both enterprise architecture and GISs. The book will then introduce a method for merging these concepts to build an enterprise GIS over an enterprise architecture foundation.

1.2 Information sources

There are numerous publications, articles, online sources, and books available about enterprise and GISs systems. This section of the chapter presents an overview of some handy resources for someone embarking on the path for building an EGIS. GIS developers are concerned with ensuring that GIS capabilities are available to everyone in an organization. Providing everyone with some GIS capabilities is the cornerstone of the fundamental enterprise GIS. The company Environmental Research Institute (ESRI), the makers of the mapping software ArcGIS, defines EGIS as a system thoroughly integrated with all levels of an organization or company and a large number of users with GIS capabilities for managing, sharing, and using spatial data (ESRI 2007a).

This definition from ESRI explains the basic concept of enterprise GIS, and it does not go into great detail, nor does it mention of considering Data

Architecture and Governance. A lack of these two items limits the definitions as source material for designing or understanding an enterprise system.

There are many online sources for information about enterprise GIS. Most of these can be passed by as unreliable. However, there is one that deserves mention, and that is the GIS Encyclopedia located at www.Wiki.GIS.com. This website has 12 pages that relate to a phase in the System Design Process situated in the section "Enterprise System Design Planning Tools." The encyclopedia does discuss system architecture design and also the importance of making sure that the GIS supports business strategy, goals, and information systems. It provides a quick overview of the design process with an overly simplified explanation of how to identify business needs and support requirements. It does finish with a nice warning that "trying to build a GIS without completing a proper system architecture design can lead to system deployment failure." I have seen several systems that have fallen victim to the warning's sentiment! Unlike other online resources, the GIS Encyclopedia is an ESRI online publication. When the manufacturer of software sets the standards in the field of GIS and the software is used by the majority of professionals, one can assume it will have useful information.

"OpenSource Enterprise GIS: An online OpenSource Enterprise GIS Manual" can be found at http://eos-gis.wikidot.com/tools is another online resource available to GIS designers. It lists pages devoted to strategy/planning, software, customization, and project management. There is a resource section that promises information on organizations and tools. Unfortunately, the site is minimal on resources and information.

Unfortunately, the OpenSource Enterprise GIS does not live up to that promise. Not one page under Chapters contains any useful or helpful information. The resource pages have links to only two organizations: The Open Source Geospatial Foundation and Codehaus; neither of which are concerned with enterprise GIS. The OpenSource Enterprise GIS is not a valuable resource for anyone wishing to learn, design, or implement a GIS.

"What's Your Definition? Looking at What Enterprise GIS Means" by Christopher Thomas is a well-written piece that discusses people's different ideas and definitions of enterprise GIS. Thomas has two critical criteria for determining when a system has matured to the enterprise level. First, the organization's system is on a centralized database that receives contributions and updates from all departments (Thomas 2009). A centralized database that is accessed by everyone in the organization is the sign of a fully integrated GIS.

The second criterion is to have support from everyone in the organization. Total support or "buy-in" by executives and staff only happens when everyone feels that GIS is a critical tool and that they can be contributors to the system's future growth, and receive valuable results from the system (Thomas 2009). When these two criteria are satisfied, the system has made the transition from a desktop or ad hoc application into a full enterprise GIS.

This article, although short, had some interesting points about the perception of what constitutes an enterprise GIS. I found this piece to be very informative, and it will help establish how to determine the threshold level for transitioning a GIS into an enterprise system. Thomas does provide a short look into how a GIS tends to grow within an organization.

R.E. Sieber discusses in his paper "GIS Implementation in the Grassroots" how an enterprise GIS develops. Sieber examines how grassroots conservation organizations implemented GISs and strategies over five years. The article points out that there are six critical factors to implementing a GIS: evaluation of need, executive support, resources dedicated to the project, training, and clear lines of communication throughout the implementation project (Sieber 2000). These are all critical areas of planning that cannot be ignored by anyone designing and implementing data or GIS architectures.

One key area that Sieber discusses is ensuring management's support for the project. If people notice that managers and supervisors are learning the new technology, they will also attempt to accept the latest technology. However, if management shows no interest, then the project will encounter resistance from those who resist change. The article has an excellent point that the coordinator must not allow management to become "bogged down" by the technology. Otherwise, executives will be surprised by the amount of change and expense involved for implementing the system (Sieber 2000). Executives must always be kept aware of the cost and amount of difference to the organization that a new system creates. When managers are surprised about a change in price, they begin to question the validity of the project and might start to withdraw their support. A lack of support could lead to a new enterprise system's implementation being delayed or canceled.

Kenneth Dueker and J. Butler in their article, "GIS-T Enterprise Data Model with Suggested Implementation Choices" examined how to build data models for enterprise-level GIS transportation systems. The article does an excellent job of laying out how a data model can be constructed to create geodatabases that support enterprise GIS. Dueker and Butler demonstrate that models are generally not seen by system users, but are critical components creating successful enterprise systems (Dueker 1997). Their data model can be applied to any information system and provides an excellent blueprint for people to track how the information will flow through the entire system.

Although the paper's audience was intended to be designers and users or enterprise GIS transportation systems, their basic premise and methodology are adaptable to other GISs. The authors promote a universal data model that uses relationships between items in the GIS (Dueker 1997). A global model brings the advantages of clear communication between clients and users, simplifies requirements for software and vendor contracts, and allows integration within the business architecture (Dueker 1997).

The white paper "Enterprise GIS for Local Government" issued by ESRI in December 2007 provides a definition and guidelines on creating an enterprise GIS. ESRI recommends that implementation of an enterprise GIS should consider existing organization standards. Methodologies, data procedures, and software applications need upgrading to versions that are compatible in the enterprise system. IT and GIS staff will need to be trained to support and work with the new system (ESRI 2007). The paper presents an excellent summary of each phase of the transition from the existing GIS into an enterprise GIS.

"Enterprise GIS for Local Government" presents a picture of the steps to be followed when an organization transition an existing GIS to an enterprise system. ESRI provides the reader with some solid ideas that are explained and clarified with some very nice graphics. Unfortunately, this paper does not address how to create a strategic plan for the enterprise GIS. It touches upon business architecture and intelligence but ultimately fails to describe how the enterprise system will support these two areas.

The publication "Best Practices: Enterprise GIS" (January 2007) provides a detailed explanation of how to design and create an enterprise GIS. A GIS professional who is overseeing the upgrade of an existing GIS into an enterprise GIS must evaluate each department's dataset for completeness, the scope of information, accuracy, where and how the data can fit into an enterprise GIS (ESRI 2007). The publication is designed to provide an overview of the design process, and it does not delve into the details of architecture design.

ESRI's "Best Practices: Enterprise GIS" provides several case studies that highlight successful enterprise GISs. "Kick-Starting Enterprise GIS" gives the impression that a reader will receive information on how to create momentum for his or her enterprise implementation project. Readers will be disappointed because the section does not contain a "how we did this project" (ESRI 2007) section. All of the case studies are summaries about each project. None of the case studies offer any solution or instructions on how to recreate the methodologies presented in the book.

1.3 Enterprise architecture

The field of GIS must be given the basic understanding of enterprise architecture if it is to adopt enterprise principles, methods, or ideas successfully. Anyone canvassing the internet will discover many websites or blogs offering slightly different versions of the same definition. Out of all of these definitions, I believe that Margaret Rouse's explanation is concise and the easiest for everyone to understand: "An enterprise architecture (EA) is a conceptual blueprint that defines the structure and operation of an organization. Enterprise architecture intends to determine how an organization can most effectively achieve its current and future objectives" (Rouse 2007). The architecture is an organization's strategic plan that determines how each company division, resource, or member can support the company's primary goal.

Rouse's online article "Enterprise Architecture (EA) Definition" is easy to read and provides a definition of enterprise architecture that can be easily understood. This article is well organized and offers several good points about the overall structure of enterprise architecture. It is part of a series of articles that can is on www.searachcio.com.

A good enterprise architecture succeeds in bringing order to a chaotic situation.

Karl Schulmeisters' article, "Enterprise Architecture: Bridging Entrepreneurs and Hard Problems" makes the point that creating an enterprise architecture means indicating a structure upon an organization that will unite resources and staff to pursue a goal (Schulmeisters 2014). A well-structured plan allows everyone to understand his/her role, responsibilities, and enable individuals to contribute to the process. Not only will this provide clear communication across departments, but it will allow the business to discover what capabilities, products, and resources are available to potential clients (Schulmeisters 2014). A company based on well-planned enterprise architecture will be able to adapt to the changes in the market and remain competitive.

Karl Schulmeisters' article, "Enterprise Architecture: Bridging Entrepreneurs and Hard Problems," is an excellent resource for anyone learning about the field of enterprise architecture. Schulmeisters avoids using the classic "buzz word" strategy or filler words found in many corporate white papers. The result is a writing style that is direct, professional, and to the point. This article was accessed and downloaded from the Orbussofware.com.

Asif Gill's article, "Defining a Facility Architecture within the Agile Enterprise Architecture Context," on Orbussoftware.com discusses how an agile enterprise is a successful enterprise architecture. This article is about how to create a "geographically distributed network" (Gill 2013) between facilities that are local or found in other countries. He introduces The Gill Framework, an enterprise architecture used in conjunction with The Open Group Architecture Framework (TOGAF) 9.1. TOGAF is a complete framework that creates a method for incorporating different systems into one overarching enterprise. The resulting system is adaptive and able to provide structure for a geographically distributed network of sites.

Gill's article, "Defining Facility Architecture within the Agile Enterprise Architecture Context," is an excellent resource about working with a geographically distributed architecture. He presented a practical methodology that has encouraged me to give his points some strong consideration. Gill has an exquisite professional style of business writing, which is enjoyable to read.

The critical principle of enterprise architecture is to coordinate integration between an organization's departments, resources, and staff. Integration and coordination will result in a coherent architecture blueprint that will establish a strong foundation for communication among everyone involved in the enterprise. Creating and maintaining clear avenues of communication are critical to the success of an enterprise system.

A company that uses an architecture that encourages everyone to communicate concerns, ideas, or suggestions freely provides a sense of user ownership, which in turn generates trust and faith in the system (Sieber 2000). When everyone supports the system, it is in harmony, and the business receives the full benefit of all resources and personnel working together to accomplish the company's business goals. When the individuals and departments are in harmony, it becomes a system that proves "the whole being greater than the sum of parts" (Michigan Department of Information Technology 2007) and provides more significant benefits to the company.

When everyone has accepted and supports the enterprise architecture, the business or organization can realize the benefits of a well-designed system. Company organization is one of the first areas to be impacted; system design, data flow, and personnel changes during takeover or a merger will be more efficient and less confusing.. The completed enterprise architecture will establish a streamlined decision process, which encourages collaboration between managers and clients, etc. (Michigan Department of Information Technology 2007).

The Gartner research team in the article, "Gartner Clarifies the Definition of the Term 'Enterprise Architecture,'" made the point that enterprise architectures should be considered a process of meeting strategic goals, not one that produces output artifacts of reports, tables, etc. (Lapkin et al. 2008). It is the job of the enterprise and business architects to keep the system design and implementation efforts focused on achieving the overall strategic purpose. There is a real danger that the project team concerned with producing "deliverables" like guidelines, standards, charts, etc., possibly causes the system to lose connection with individuals or departments (Lapkin et al. 2008). Getting stuck would result in a counterproductive situation because people or departments would feel ignored or overlooked and not support or participate in enterprise architecture. This book has stated that systems without everyone's participation would eventually fail.

Clear and concise communication about the architecture's requirements, the overall goal between everyone involved in the design process is critical to the project's success. Louw Labuschagne's white paper "Building Enterprise Architectures for Non-Architects" published on Orbussoftware.com is an excellent explanation of enterprise architecture and the need for clear communication. Communication is the key to enterprise architecture's success or failure. Enterprise architecture is about managing the company's data, risk, and response to change (Labuschagne 2011). A manager must ensure that all team members understand their roles and responsibilities on the project. When everyone understands their role in the system, the exchange of ideas and compromises can flow among the segments of the architecture plan. Management of a company's risk and response to change happens with effective communication between people.

1.4 Conclusion

Enterprise GISs are primarily designed and implemented to provide GIS capabilities to departments and members of an organization. The papers and publications often provide a summary of the processes needed to develop and distribute a GIS but give very little "how-to" details of system design. Unfortunately, GIS professionals, developers, and designers have failed to provide these systems with a strategic goal or purpose. Each person or department produces maps and data with no sense of organization or mutual project support. This ad hoc distributed GIS results in duplication of efforts, data, and results. There is chaos within the system as people produce large amounts of information without a sense of direction or value.

The literature demonstrates that enterprise architecture is about providing organization, managing risk, and change within a complex system. An organization's architecture succeeds only if it achieves the total support or "buy-in" from departments and members (Wilson 2013). Communication is a crucial element that enables the system to manage risk and respond to change.

It is disturbing that the online resources and publications available to GIS professionals and users do not explain the details involved in creating enterprise systems. ESRI laid claim to developing enterprise architecture but provided no detailed information for any phase of the design process (Wilson 2013). GIS literature does not provide a discussion about applying data standards and principles, or ensuring that the system supports the organization's goals. A reader can quickly develop the idea that a successful enterprise system is one that provides a GIS toolset across all levels of an organization.

The enterprise GIS provides tools to all levels of an organization, but such a system lacks organization and direction. Therefore it will not be effective in supporting the company's business goals. The discipline of enterprise architecture can provide all of those features that are missing from the enterprise GIS. Adding the theories, methodologies, and visions of enterprise architecture to enterprise GIS will align the system to support the organization's goals. A skillful merger of enterprise architecture and GIS will bring order to a system in chaos.

1.4.1 Organization

The chapters in this book will guide the reader through the process of building the enterprise GIS over a foundation of enterprise architecture principles and methodologies. It is essential to understand the similarities and differences between enterprise GIS and enterprise architecture. Therefore Chapters two and three are quick introductions to the fields of enterprise GIS and enterprise architecture. Chapter four will introduce and define the various positions and roles that are required to establish an enterprise architecture and then move on to explain the counterparts in enterprise GIS. Chapter five will introduce and discuss the realm of data architecture and how this architecture works within a GIS.

Chapter six is an introduction to various tools and methodologies that the author has been successful in using during enterprise system builds. A successful methodology for incorporating enterprise architecture into an enterprise GIS will be covered by Chapter seven.

Chapter eight teaches students how to visualize the new information system. Readers learn how to look for connections between systems and determine the flow of information between areas of the system. The final part of Chapter eight discusses how to use these techniques to visualize the future system.

Chapter nine discusses the importance of system governance, or management as a defense against data corruption, system chaos, and pursuing tangent or dead-end paths for development. The chapter introduces the concept of a review board that ensures enforcement of all the policies, standards, and methodologies contained in the architecture policy.

Chapter ten brings the book to a conclusion and presents a short discussion about the future of GIS, as well as how it can work with big data and develop a global view. The last part of the chapter offers a few thoughts about potential lines of research open to those who wish to explore further merging enterprise architecture and enterprise GIS.

Students will enhance their knowledge and skill sets by completing the chapter and concept assignments within this book. These assignments will provide an introduction to many issues that can challenge the successful design and implementation of enterprise information systems. Each assignment will deal with a specific concept introduced by the text and allow readers to experience first hand the process of creating an enterprise system.

1.5 Applied GIS

Many fields outside of geography have found a use for geographic information systems (GISs). Historians have successfully created a digital version of the Battle of Little Bighorn for real-time analysis by creating an artifact location database and then mapping the battle's progress with GIS. Public Health and Emergency Management officials have used GIS to predict, track, and respond to disease outbreaks around the world. GIS has become a valuable tool to government, industry, and business leaders.

The growth and development of many GISs have followed an ad hoc pattern. One or two people will start working with GIS software to produce maps for a small project. The success of these maps will encourage other people to start working with GIS. Eventually, there are several people or departments using GIS, but with little or no coordination. The idea of the CIS is to share files and information; it lacks coordination. Each person or department follows their path with little regard for others or coordinating their efforts to produce a coherent system. Many people would consider this ad hoc system a successful implementation of GIS within an organization; in reality, it is chaotic. It is merely a system of chaos and disorder where each GIS user follows his or her path. GIS professionals should use the ideals of enterprise architecture to provide organization, vision and principles that bring a sense of order to a chaotic system.

Anyone attempting to bring order to a chaotic GIS must understand the similarities and differences between enterprise GIS and enterprise architecture. This book expresses the belief that enterprise GIS can benefit from the theories and methodologies of enterprise architecture. This book does not advocate "scrapping" enterprise GISs but explains how the two systems can coexist, support, and work together. This author believes that the ideas and strategies of enterprise architecture can enable enterprise GISs to reach their full potential.

1.6 Assignment

1.6.1 Assignment 1 Article Review

Find one article each about enterprise architecture and enterprise GIS. Read each article and write a summary and participate in a class discussion about enterprise architecture and enterprise GIS.

Bibliography

Dueker, Kenneth, J., J. Allison Butler. 1997. *GIS-T Enterprise Data Model with Suggested Implementation Choices*. Portland, OR: Center for Urban Studies, School of Urban and Public Affairs, Portland State University, October 1.

ESRI. 2007a. *Enterprise GIS for Local Government*. Redlands, CA, December. http://www.esri.com.

ESRI. 2007b. *GIS Best Practices Enterprise GIS*. Edited by ESRI. ESRI PRESS. Redlands, CA, January 1. http://www.esri.com.

Gartner, Inc. 2008. "Gartner Clarifies the Definition of the Term 'Enterprise Architecture'." *Gartner Research* (Gartner), 15. doi:G00156559.

Gill, Asif. 2013. "Defining a Facility Architecture within the Agile Enterprise Architecture Context." *Orbus Software*. Orbus Software. October 1. doi:WP0107.

Labuschagne, Louw. 2011. "Building Enterprise Architectures for Non-Architects." *Orbus Software*. September 1. doi:WP0011.

Lapkin, Anne, Philip Allega, Brian Burke, Betsy Burton, R. Scott Bittler, Robert A. Handler, Greta James. 2008. "Gartner Clarifies the Definition of the Term 'Enterprise Architect'." *Gartner*, August 12: 1–5.

Michigan Department of Information Technology. 2007. *From Vision to Action: Enterprise Architecture – Strategic Approach*. Michigan Department of Information Technology.

Rouse, Margaret. 2007. "Enterprise Architecture (EA) Definition." *TechTarget*. June 01. Accessed October 2015. http://searchcio.techtarget.com.

Schulmeisters, Kar. 2014. "Enterprise Architecture: Bridging Entrepreneurs and Hard Problems." *Orbus Software*. November 1. doi:WP0168.

Sieber, R.E. 2000. "GIS Implementation in the Grassroots." *URISA Journal* (Vol. 12, No. 1): 15–29.

The Open Group. 1995–2015. *Enterprise*. January 1. www.opengroup.org/subjectareas/enterprise.

Thomas, Christopher. 2009. "What's Your Definition? Looking at What Enterprise GIS Really Means." *ArcUser*, 3032. http://www.esri.com.

Wilson, North Carolina. 2013. A GIS Manager's dialogue: Is your business semi-integrated or fully system integrated? *North Carolina GIS Conference*, Raleigh, NC.

2

Enterprise Geographic Information System (Enterprise GIS)

This chapter will discuss how a simple GIS desktop application becomes an enterprise system. However, these ad hoc systems often lack a sense of direction or organization. This chapter ends with an examination of various sources.

What is an enterprise geographic information system (enterprise GIS)? How does an enterprise GIS develop and grow? We can use the morning glory vine as an analogy for how a geographic information system (GIS) begins to grow and wind its tendrils into every nook of an organization. It is unnoticed at first, and quietly sprouts when one person starts working with a small "experimental" GIS project on his or her desktop. The successful bloom of this first project will lead to requests from other people for GIS support. It is not long before that one lone desktop system has sprouted from one small seed project into a complex information system whose tendrils weave through every department. The result of this organization-wide GIS is a rapid increase in data, mapping, and information use.

When a GIS reaches the point of being available to anyone within an organization, it is considered an enterprise GIS. The question is, "What is an enterprise GIS?" The discipline of GIS is relatively new, and it is a virtual newcomer to concepts of enterprise databases and systems. Therefore, there are many different definitions of what constitutes or what defines an enterprise GIS.

A system is considered an enterprise when it connects and exists on all levels of an organization and allows many users to access GIS capabilities. The enterprise GIS will also support multiple data transactions, seamlessly integrate with existing systems, and follow The Geospatial Consortium (OGC) standards. It must also ensure a consistent display and sharing of information and functionality across desktop, mobile, or web-based platforms. One can expand upon this definition with Christopher Thomas' article, "What's Your Definition? Looking at What Enterprise GIS Means." This article presents two essential criteria for determining when a system has reached the enterprise level. An organization's GIS, should be based on a centralized database that allows the system to receive new data, edits, updates, and projects from all departments within the organization (Thomas 2009), the first criterion has been satisfied. A centralized database is vital because it sharply reduces the potential for data errors of redundancy, duplication, and inconsistent formats. One source of data for all analysis and mapping projects will enhance the system's performance and ability to deliver information in a quick and efficient manner.

The second criterion is organizational support for the system. No matter how big or agile the enterprise GIS, if no one uses it, it fails. Total support or "buy-in" by executives and staff only happens when everyone feels that GIS is a critical tool and that they can be contributors to the system's future growth and receive valuable results from the system (Thomas 2009). When these two criteria are satisfied, the system has made the transition from a desktop or ad hoc application into a full enterprise GIS.

The Environmental Systems Research Institute (ESRI) is a company that produces the ArcGIS suite, which has become the software standard for GIS. ESRI's software applications have become the standard for GIS software and definitions; the GIS community readily accepts practices. ESRI states that it is "[a] geographic information system that is integrated through an entire organization so that a large number of users can manage, share, and use spatial data and related information to address a variety of needs, including data creation, modification, visualization, analysis, and dissemination" (ESRI 2015). The enterprise GIS expands to include the functions of information creation, editing, analysis, and distribution.

2.1 Just what is enterprise GIS?

When one attempts to pin down what constitutes an enterprise GIS, two questions emerge. Why are there so many different, yet similar, definitions for an enterprise GIS? Where is the reference literature that should help to establish when a GIS transitions from a desktop to an enterprise system? The answer to both questions is that the reference material is on the way. We must remember that GIS is a new discipline and is still establishing a foundation of discipline definitions, parameters, and principles. It is time for users and professionals to contribute ideas and concepts to grow the foundation of GIS knowledge.

2.2 Why are there so many different, yet similar, definitions for enterprise GIS?

People in government, industry, and academia have wondered about the number of similar yet, different definitions of what is an enterprise GIS. Software vendors are one cause for this state of competing definitions.

Vendors try to convince the market that their application can "do it all"! Consider this example: Company Alpha sells a software application that is designed to track the amount of supplies (fuel, oil, parts) consumed by the equipment division. A salesman overhears a director asking about a software package that enables him to follow how office supplies are being consumed within his division. The salesman quickly rewrites the sales materials to detail how Alpha's software will manage the tracking of supplies for the office and equipment. Supporting two departments for the low cost of a single program! But wait, there is more!

Down in marketing, staff quickly realizes that the software is providing benefits to two departments, behold! They quickly remarket the application as an enterprise logistics solution and rush the new materials over to sales. Almost overnight, this simple program has transformed into a "powerful enterprise" solution.

GIS is a wonderful tool for supporting almost any type of information need or analysis. GISs are very adaptable and are applied to a wide range of situations. Due to the variety of ways a GIS can be used, it is natural for GIS programs to appear as "out of the box" enterprise solutions. Vendors quickly labeled their GIS products as "enterprise solutions" to better position themselves for product sales. The sales teams are quick to discuss how their product is an easily installed off-the-shelf enterprise GIS.

Naturally, each sales department in an effort to "separate themselves from the pack" will develop a slightly different definition of what constitutes an enterprise system. All these definitions agree that the enterprise system supports more than one department in a municipality or business. Vendors never seem to offer methods or insights on how to build the enterprise system that would integrate with the software application. A properly designed, implemented, and enforced enterprise architecture is more important than the software.

2.3 Case 1: Medina County Health Department, Medina County, Ohio

Medina County Health Department (MCHD) environmental division required a GIS for tracking mosquito response, sanitary concerns, and problem septic systems. Several internal discussions between staff, department heads, and the Health Commissioner led to the conclusion that the system would eventually provide capabilities to the nursing, community health, and administration divisions. The requirement that the system provide for future expansion set the department squarely on the path of enterprise development. The vendor who won the original contract never held a meeting to determine the needs or expectations of the health department. No assessment was performed to determine the strengths and weaknesses of the department's current systems. The oddest and undefendable error committed by the vendor was not discussing the new system with health department directors and supervisors. This vendor made the drastic assumption that their expectations and goals were the same as the health department.

MCHD staff wanted to generate reports with maps from the system. This way the information about sites, inspections, or facilities could be archived and emailed to people. Each item displayed in the enterprise GIS would be hyperlinked to a file or individual reports. The vendor never realized this requirement and their software package did not have a save or print to PDF option. Unfortunately, the proposed solution to this deficiency was to purchase a third-party software and develop a bridge between the two programs. These two extra services generated additional costs against the MCHD's limited budget.

This situation was unacceptable to the Director of Environmental Health, who reached out to the author's former employer Chagrin Valley Engineering (CVE). CVE's first action was to review all the documentation (including the original proposal) and purchases for the MCHD enterprise GIS project. This assessment provided CVE with a complete overview of the project's direction and progress.

A project review also revealed two important errors. The first mistake was a lack of clear communication between the health department and the original consultant staff. The original consultant's staff had never explained to MCHD personnel what an enterprise system was and the processes required for building or deploying one. MCHD personnel had the desire to learn and assume ownership of a new system, but completely lacked a starting point. Apparently, the vendor assumed that everyone had a basic knowledge of GIS, data collection, and enterprise systems.

Chagrin Valley Engineering's principal partners gave the author (he was their GIS coordinator) complete freedom and support to guide the health department through the process of developing an enterprise GIS. The trust and support of CVE's ownership was a critical component to the eventual success of the project. MCHD's environmental director also provided his cooperation and support to the author's efforts. This combined support not only provided instant credibility to the author's credentials and abilities but demonstrated that CVE and MCHD leadership were committed to successfully completing the project. Ultimately, the success of an enterprise system build has a direct correlation to the amount of support and commitment exhibited by management.

The author conducted several "fact-finding" meetings with MCHD supervisors, staff, and management. Each meeting was designed to determine what exactly were the perceived needs, goals, and expectations each group held for a GIS. These meetings allowed the airing of disappointments about the previous consultants, while providing a platform that stimulated new ideas and excitement for the project. These fact-finding meetings provided proof that everyone's ideas or concerns would be considered by the team at CVE.

These meetings resulted in a strategic plan for developing an enterprise GIS for the MCHD. Environmental Health's mosquito control team would be the first ones provided with GIS capabilities, and other departments would be added when requested. A supervisor would be trained and mentored in GIS management, while a member of the field crew would be provided training as a GIS technician. This split in responsibilities would allow the MCHD staff members to learn GIS procedures without adding responsibilities outside their paygrade responsibilities. This plan for splitting the learning responsibilities of a GIS was the perfect solution for the situation within the MCHD.

CVE's ownership supported the project with a unique contract. A two-year contract was provided to the MCHD with the first year being a standard to build, implement, and provide a training project. The unique piece of the contract occurred in the contract's second year. CVE's geospatial team transitions from the role of builder/training into a mentor/troubleshooter. This unique approach to an enterprise build contract allowed the health department to build and deploy a GIS, and then have professional support for problems that were outside the scope of their newly acquired skill sets. CVE's contract with MCHD provided health department staff the confidence to attempt problem solving on their own, with the knowledge that they could easily call and quickly receive technical support.

This contract established a relationship between CVE and the MCHD that was more along the lines of a partnership or coach/mentor, rather than a traditional contractor and client. The project accomplished the development and implementation of a GIS within the first year of a two-year contract. This includes the training of an MCHD GIS manager and technician to provide data mapping and analysis to all the health department's divisions. Six months into the second year, the MCHD was producing and presenting maps for projects, county citizens, and conferences. Due to this success MCHD and CVE agreed not to use the option for a third year, and CVE simply made the transition to a "call when needed" consultant.

2.4 A word on professionalism

MCHD personnel had three reactions to the original contractor's assumption: anger, dismissal, and "whatever." The anger came from the belief that MCHD had been "ripped-off"; it had paid for a GIS and field data collection equipment that were simply stuffed in a box gathering dust. A lot of money that should have been used for new equipment, supplies, or personnel had been spent on a box of junk. The impression of being ripped-off was the direct result of the lack of communication between the MCHD and the original contractor. Anger is a normal reaction when someone perceives that they have been cheated or shortchanged by a supplier.

Dismissal is an aftereffect of a failed project and feelings of abandonment when the contractor leaves a dissatisfied client behind. Any initiative for resurrecting the old or beginning a new project is almost immediately rejected on the basis that it has already been attempted and failed. The attitude of dismissal is extremely frustrating to encounter, because the client's staff simply does not want to listen. This attitude can be overcome with patience and a little effort.

A consultant's professionalism is one of the biggest weapons to combat a dismissive attitude in a client's staff. Everyone should realize that professionalism consists of being on time, having a clean appearance, properly prepared for work, keeping promises, always giving a full effort, knowing personal limits, and being honest. Unfortunately, not everyone understands the value of being on time. Walking into a client's office even five minutes late creates a negative impression with the staff. One should always remember that the staff must be there on time, otherwise their pay can be docked. Any staff who must delay the start of their work, because of a habitual tardiness will quickly begin to resent the consultant.

Resentment will only build if the consultant staff underdresses for the client's business environment. A person who walks into a business wearing a t-shirt, cargo shorts, and sunglasses will lose respect. Consider a situation where your car has developed a knocking sound. Two men offer to help you determine where the sound originates. The first man is wearing a $300.00 suit and tie, and the second one is wearing greasy coveralls, with a wrench in the back pocket. Who would you assume to be the expert mechanic? More than likely, the reader will join the majority and listen to the man in greasy overalls. Because he looks like a knowledgeable mechanic, despite the fact that there is not a service garage in sight. People respect and pay attention to someone who is dressed the part of a knowledgeable and skilled professional.

The author's personal experience confirms looking the part that often sells presenter as an expert. He spent 10 years in the U.S. Navy Reserves with the Seabees (Navy Mobile Construction Battalion NMCB – CB = Seabees) as an equipment operator, leader, and instructor. Seabee instructor training drove home the point that an effective teacher was the most professional-looking person in the room. This means a fully pressed uniform with military creases, boots to a highly polished shine, and a proper sharp edge military haircut. Experienced navy instructors taught the point that if you look the part, then you will have more confidence in yourself. This self-confidence is picked up by the audience and reinforces your image as a knowledgeable professional. Looking professional helps your confidence while reinforcing the image of a skilled and knowledgeable professional.

Being prepared is the motto for the Boy Scouts of America (BSA). It should be the stand for anyone entering a business or organization as a professional consultant. First impressions are critical in the world of consulting. Anyone who enters a client's workspace and has to "lookup" answers to questions is not going to gain the confidence of the client's staff. People can tell when someone does not know what they are talking about. One should never walk into a situation without a full understanding of the project parameters or history. When a person has the knowledge to answer questions about projects, programs, or systems without the need to consider, research, or defer to another, it cements the client's trust in their advice. A client's trust is a valuable commodity for success.

Trust is a crucial piece of the relationship between a client and the consultant. A professional appearance, attitude, and knowledge will quickly establish a tentative trusting relationship with the client. The way to move from a tentative to a firm trusting relationship is by living up to one's word or better known as keeping a promise. Every project begins with a clean slate of trust, but missed deadlines, excuses, program shortcomings, misinformation, will chip away at this clean slate. Eventually the broken promises result in the client terminating the consultant's contract or the project and finding a new consultant.

A good consultant will lead by example! Don't walk away when there is a problem, roll up your sleeves, get down, and get dirty to help solve the problem. One of the worst habits this author has witnessed is a "professional" who states, "oh it is 4:00 pm, you guys finish up" and walks away. The best way for a consultant to win the trust and support of a client's staff is by leading through example. When a problem surfaces (and they will happen), the contractor should be the first person going the extra mile: staying late, buying snacks for those working overtime, calling fellow professionals for advice, etc. Become part of the team, finding solutions to problems. The perception that you not only sympathize but are willing to roll up your sleeves and "get dirty" to help them solve a problem is invaluable in gaining the client's team respect and attention. When client is struggling with system problems, working past normal working hours, losing time from family, a consultant who is willing to work only an eight-hour day is not winning respect, friends, or new clients.

An important aspect of consulting or system design can be summed up in the phrase "Know your limitations." One of the first lessons taught to new Petty Officers in the U.S. Navy is that it is ok to say "I don't know the answer," as long as you follow that statement with "I will find the answer or find someone who can answer your question." It is ok to admit that one has reached the limits of their own knowledge and that someone else can provide the answer. Clients will have more respect and trust in a consultant who is not afraid to admit to the limits of their knowledge or skills, especially, when the person knows whom they can reach out to for support and the proper answer.

The author has had the misfortune to follow in the footsteps of consultants who did not follow the rules of professionalism. He had to deal with and overcome a client's feelings of anger, dismissal, and "whatever" that were left behind by unprofessional consultants. His solution is to be ultra-professional, friendly, and honest. Professionalism does not require being cold and reserved to the point of arrogance. A person can be professional while sharing smiles, greetings, laughs, with people. An honest, friendly, smiling person will eventually regain the trust and respect that was lost by the previous consultant's poor performance.

A professional is always honest with their client. The author does not mean being honest about what one might think about the client's appearance, office paint scheme, etc. These are personal items that should not impact how a consultant performs his work. The professional must be honest when discussing work items, costs or expenses, time requirements, and explaining why a certain course of action might not be the best idea. Never use the phrase "I am just telling you the truth"; this line is a simple cop out used by people who want to make a criticism and not be challenged by their target. Remember, an honest opinion must be backed up by good solid reasoning. Clients will always appreciate an honest professional, who attempts to guide, mentor, or help them grow the business.

Honesty also includes being transparent about your work. When working on a project, explain the why and how of each step that you or the team are completing. An honest consult does not need to hide or shield part of the project from questions or scrutiny. The minute a question is dodged, vaguely answered, or ignored, people's trust in your word will be at the very least damaged if not lost. When someone does not understand what is happening, be honest and provide a reason why the team must perform this step and doing these steps in this order will result in reaching a project goal. Sometimes, a professional will act as a mentor to the client and their staff, coaching, and teaching about the processes involved in developing enterprise systems.

The mistake of assuming that your goals and expectations match or will be accepted by the client is the first step on the road to disaster. A successful enterprise, information, or any system meets or exceeds the expectations and needs of the client. One should never make the mistake that they know what is best for a client. The basic truth of consulting and system design is that only the customer knows what is best for their organization.

When designing systems, this author cannot overstate the importance of understanding and meeting the client's expectations! Remember, once the system is installed the client must live with it, not the vendor. Any problems or issues will fall at the feet (or desk) of the person(s) who contracted and authorized the work. A successful consultant or designer always ensures that the new system will incorporate and address the client's concerns. Additional charges for software and programming should have been avoided with a simple conversation with the health department's environmental director.

2.5 The enterprise GIS

The purpose of this book is to add a few more stones to this knowledge foundation. First, let's assemble the components of the definitions, as mentioned earlier into one. A GIS transitions into an enterprise system when the following conditions have been met: 1. It is based and operates from a centralized spatial database. 2. The system connects multiple departments with multiple users. 3. The organization's executives, managers, supervisors, and staff support and feel they have an ownership stake in the system. 4. It provides these departments and users with a standard set of tools for the creation, editing, analyzing, mapping, and distribution of information.

Although this definition incorporates the crucial elements from the previous versions, it is not complete. There are several elements of the enterprise system that are missing. This GIS enterprise definition does not provide any form of organization strategy, principles, standards, or even goals. The discipline of enterprise architecture can deliver these elements to GIS.

The business world has demonstrated that a common practice is to hire a person with GIS experience and then expect them to build a fully functional system. Once the new GIS person has become established and has developed a desktop system, then system expansion begins with the question of "How do we take the next step and move into an enterprise system?" The real answer to this question is that moving from a desktop to an enterprise system will be a lengthy process.

There are several essential items to consider when discussing the implementation of an enterprise GIS.

- The GIS supports the organization's business strategy.
- The GIS supports the business goals.
- The GIS will integrate with internal and outside information or business systems.

    The GIS will provide benefits to more than one layer, division, or department of the organization.

Considering these items will help the designer determine the best fit for the system to the organization. The reason for developing an enterprise system is to ensure that the company grows. This requires an information system that provides good information, which will enable decision-makers to plan strategies, goals, and markets for the organization. A system that does not support the business strategy or goals will not help the business' long-term growth.

Communication between organization members and/or clients is a critical component of success. The old way of communicating consisted of letters, phone calls, or a face-to-face meeting between people. Our digital world connects business or people almost instantaneously via email, video conferencing (webex, joinme, etc.), online chats (facetime, wechat, etc.). Today's digital connections enable a business to stay local, yet tap into a global network of suppliers, customers, and opportunities. The ability to connect, accept, or export information, across different networks or platforms is key to a successful business. Potential clients or opportunities will be lost, if there is the perception that the company cannot connect with outside systems or data sources.

A new system must provide benefits not only to the original department but to other divisions, departments, or sections of the company. A system can only be considered a true enterprise if it provides capabilities and advantages to every level of the business. This is a key concept of enterprise projects, that the project must have a benefit to more than just one department. However, the new enterprise system must not only provide benefits to all levels of the organization but be agile, flexible, and able to adapt to changes within the digital world. A flexible system will support and enable a business to exploit new opportunities that appear in the digital world.

2.6 The beginning of the enterprise system

Where does one begin? The first step would be a complete evaluation of the organization's GIS structure, data, and goals as recommended in the ESRI publication "Best Practices: Enterprise GIS," published in January 2007. This process will take time and effort because each piece of the GIS must be carefully evaluated to determine if there is a need to move into an enterprise GIS. The evaluation will entail a complete review of the existing GIS structure, resources, and requirements to determine how a reorganization or expansion of the current system will impact its ability to support the organization's current or future goals and business practices.

An assessment of all existing GIS resources (staff, data/information, projects, support requests by department) when compared to the growth of GIS-related projects within the organization will aid in determining the current and potential need of an enterprise GIS. The assessment should include the following questions: How many people are using GIS capabilities? What are the number of projects and departments to be supported? Is the current technology level of the computers, servers, and network capable of supporting an enterprise GIS? Determining and being able to prove the organization's growing "need" for an expanded and robust GIS is critical in gaining support for an implementation project (Sieber 2000).

The GIS assessment will also have to provide an inventory of the resources which currently are committed to supporting the system and how many resources will be consumed by the implementation and maintenance of an enterprise system (Sieber 2000). A complete evaluation and inventory of the resources will result in a cost estimate for the entire upgrade/implementation project.

A report that covers the how, why, and what benefits an enterprise GIS will provide to the organization is critical toward gaining support from management. Not only will the analysis show due diligence on behalf of the GIS staff, but its professional, clean, and assessment results can be placed into an easy-to-understand format that will be familiar to business leaders. It is also a solid foundation for a discussion defending and explaining the need for a project to upgrade from a desktop to an enterprise GIS. It is critical to work with methods that are familiar to management so they can easily understand how the organization will benefit from the enterprise GIS. It is essential that executive management understands, accepts, and supports the enterprise GIS (Sieber 2000).

Assessments are a critical component for determining what resources the organization has at its disposal. Assessments often reveal items that management were not aware of (pilot projects, datasets, software, etc.) and an honest overview of the organization's IT infrastructure (age, upgrades, capacity, etc.). This assessment information will be used to develop a strategic plan that will guide the organization's enterprise build project.

Consider the situation presented in case 1. The Medina County Health Department desired an enterprise GIS but had no concrete plan for building and implementing a GIS. Chagrin Valley Engineering performed a detailed assessment that determined what internal resources were available, resources that were lacking, personnel skillsets, IT infrastructure, and a plan for IT upgrades and acquiring required resources. CVE's assessment became an important part of the MCHD's strategic plan for developing, implementing, and maintaining an enterprise GIS.

When people notice that managers and supervisors are trying to learn new technology, then they will also attempt to accept the latest technology. However, if management shows no interest, then the project will encounter resistance from those who resist change (Sieber 2000). Maintaining clear lines of communication is also vital in keeping the support of executive management (Sieber 2000). Executives must always be kept aware of the cost and amount of change to the organization that a new system creates. Any manager blindsided about an increase in price; they begin to question the validity of the project and might start to withdraw their support. Loss of the manager's support could lead to a new enterprise system's implementation being delayed or canceled.

One cannot overstate the importance of project support by ownership, executive management, and supervisors. Support from these levels in an organization will have a positive impact on how the employees or staff view any new project. When ownership or executive management shows that they consider a project important, most of the general staff will express a similar viewpoint. Ownership and executive management are the decision-makers for the organization, the staff will look toward these people for guidance. Generally, once the organization's leaders accept a project or idea, the employees will also accept it.

Among humans, there will always be a group who will resist change. These people will offer up all sorts of reasons why they can't accept this new project. Despite all their protests and excuses their resistance is really based on fear. Fear is a result of miscommunication, a higher tenure, and changes in routine. Miscommunication concerning the new enterprise system's purpose will lead to a fear that certain positions (people) will be found redundant, unnecessary, obsolete, and these people will be terminated. Older employees often feel that they are past the age of learning how to work with new technologies. Long-time employees tend to find comfort in a daily routine and the predictability of those in management.

These high-tenured employees might have problems learning a new information system or adjusting to changes within the office. Older employees fear that any problem learning a new system or technology will result in their termination from the company. Everyone should remember that these long-time employees are a valuable asset to the organization's new information system. A person who has enjoyed a 20- or 30-year career within the company has acquired historical knowledge of business operations, decisions, and procedures that are not found in the company archives. This "old-timer" knowledge can prevent the team designing and implementing a new enterprise system from incorporating past mistakes into the digital data processes. Management and the consulting team must ensure that the valuable resource of old-timer knowledge is not lost by employees' terminations or resignations. Company old timers should be encouraged to participate and share their knowledge with the design team.

People overlook doing a close evaluation of the organization's current GIS data, information, and practices. Accurate data is the key to any GIS desktop or enterprise system. Data errors like typos, misspellings, improper whitespaces in names, etc. must be discovered and removed. An assessment of the GIS processes reveals redundant, duplicated, or unnecessary processes which, when eliminated, will increase system efficiency (ESRI 2007b). One data failure at a critical moment can result in everyone losing faith in the GIS's ability to produce accurate and reliable results.

During the datasets evaluation, work can begin a data model. Kenneth Dueker and J. Butler in their paper "GIS-T Enterprise Data Model with Suggested Implementation Choices" examined how to build data models for enterprise-level GIS transportation systems. The authors demonstrate that models are generally not seen by system users, but they are critical components creating successful enterprise systems. This concept of a data model can be applied to any information system and will provide a blueprint (or chart of pathways) about how the information will flow through the system.

Dueker and Butler's writing targets enterprise GIS transportation systems; the basic premise and methodology adaptable for non-transportation GISs.

The authors promote the idea of a universal data model based on relationships between items in the GIS (Dueker and Butler 1997). This type of model brings the advantages of clear communication between clients and users, simplifies requirements for software and vendor contracts, and allows integration of the GIS into the business architecture (Dueker and Butler 1997). Seamless integration between systems will create an accurate, efficient, and agile information system that will provide many long-term benefits to the organization.

ESRI's ArcGIS software and practices are considered industry standards. ESRI recommends that the implementation of an enterprise GIS should consider existing organization standards. Methodologies, data procedures, and software applications receive an upgrade to versions that can integrate into an enterprise system. IT and GIS staff will need to be trained to support and work with the new system (ESRI 2007). Establishing standards, creating consistent methodologies, and procedures will streamline the enterprise GIS.

GIS professionals, users, and system designers are still learning how to create, deploy, or work with the concepts of enterprise systems. There is much discussion in the GIS about establishing and using an enterprise system. However, the literature in the field does not provide any detail on what is or how to create, implement, or maintain an enterprise GIS. None of the GIS experts or websites offer a detailed "how-to" approach to create or convert a desktop into an enterprise system. Instead, the perception is that if a GIS has spread from one desktop to many, then an enterprise system has been successfully established. However, this type of system has no sense of order; instead, it encourages chaos as everyone takes a different path with no regard for others. The result is an information system that produces large amounts of data but is unable to fully exploit this data or turn it into information that provides value to the organization.

2.7 Summary

GIS professionals, users, and developers need to understand that a successful enterprise system has a strategic vision. This vision becomes the blueprint that binds the system together and focuses all resources, staff, and projects into one team that supports the organization's business vision and goals. The professionals of GIS can incorporate and use the ideals of enterprise architecture to improve and enhance the value of their enterprise systems.

2.8 Assignment

Assignment 2: Is there a benefit?

A small company that employs 30 people in the administration, sales, engineering, environmental sciences, and surveying would like to implement an enterprise GIS.

Why should this company consider implementing an enterprise GIS?

Would this company benefit from an enterprise system? Yes or No. Please explain your answer.

Bibliography

Dueker, Kenneth, J., J. Allison Butler. 1997. *GIS-T Enterprise Data Model with Suggested Implementation Choices*. Portland, OR: Center for Urban Studies, School of Urban and Public Affairs, Portland State University, October 1.

ESRI. 2007a. *Enterprise GIS for Local Government*. Redlands, CA, December. http://www.esri.com.

ESRI. 2007b. *GIS Best Practices Enterprise GIS*. Edited by ESRI. ESRI PRESS. Redlands, CA, January 1. http://www.esri.com.

Gartner, Inc. 2008. "Gartner Clarifies the Definition of the Term 'Enterprise Architecture'." *Gartner Research* (Gartner), 15. doi:G00156559.

Gill, Asif. 2013. "Defining a Facility Architecture within the Agile Enterprise Architecture Context." *Orbus Software*. October 1. doi:WP0107.

Labuschagne, Louw. 2011. "Building Enterprise Architectures for Non-Architects." *Orbus Software*. September 1. doi:WP0011.

Lapkin, Anne, Philip Allega, Brian Burke, Betsy Burton, R. Scott Bittler, Robert A. Handler, Greta James. 2008. "Gartner Clarifies the Definition of the Term 'Enterprise Architect'." *Gartner*, August 12: 1–5.

Michigan Department of Information Technology. 2007. *From Vision to Action: Enterprise Architecture – Strategic Approach*. Michigan Department of Information Technology.

Rouse, Margaret. 2007. "Enterprise Architecture (EA) Definition." *TechTarget*. June 01. Accessed October 2015. http://searchcio.techtarget.com.

Schulmeisters, Kar. 2014. "Enterprise Architecture: Bridging Entrepreneurs and Hard Problems." *Orbus Software*. November 1. doi:WP0168.

Sieber, R.E. 2000. "GIS Implementation in the Grassroots." *URSIA Journal*. Vol. 12, No. 1: 15–29.

The Open Group. 1995–2015. *Enterprise*. January 1. www.opengroup.org/subjectareas/enterprise.

Thomas, Christopher. 2009. "What's Your Definition? Looking at What Enterprise GIS Really Means." *ArcUser*, 3032. http://www.esri.com.

Wilson, North Carolina. 2013. A GIS manager's dialogue: Is your business semi-integrated or fully system integrated? *North Carolina GIS Conference*, Raleigh, NC.

3

Enterprise Architecture

3.1 What is enterprise architecture?

To successfully use enterprise methods or ideas, a basic understanding of the principles of enterprise architecture (EA) is necessary. While there are many variations on the definition of EA, we will use Margaret Rouse's definition: "An enterprise architecture is a conceptual blueprint that defines the structure and operation of an organization. Enterprise architecture intends to determine how an organization can most effectively achieve its current and future objectives" (Rouse 2007). The architecture is an organization's strategic plan that determines how each company division, resource, or member can support the company's primary goal.

Good EA works and will succeed in bringing order to the chaos of daily business operations. Karl Schulmeisters in the paper "Enterprise Architecture: Bridging Entrepreneurs and Hard Problems" makes the point that creating an EA equates to creating a structure around which an organization will unite resources and staff to pursue a goal (Schulmeisters 2014). It is the enterprise architect's job to lay down the proverbial law in the form of a well-structured plan that informs everyone of their roles and responsibilities within the organization. A clear-cut, concise, and easy-to-understand explanation of how their efforts benefit the organization will encourage buy-in from staff at every level of the organization.

Gaining the support of an organization's staff is important to the success of any enterprise project. Friendly and supportive staff will steer the consulting team away from potential in-house political pitfalls or social faux pas that might derail the new project. The author was fortunate enough to benefit from staff support during his time at the Summit County Health Department. A supportive worker tipped the author that the IT supervisor had heard she would be expected to lose staff because the new GIS project required fewer support hours. Naturally, she was upset and wished to defend the IT department from a perceived threat of a cut in staffing levels.

Responding to the tip, this book's author switched from an opening statement about what capabilities a GIS could offer IT, to the department might like to consider adding staff and requesting a larger budget for supporting the new system. This opening statement caught the supervisor by surprise, transformed her attitude from combative to curious, and provided a space for a conversation about how the IT department could support the enterprise project. The author was able to exploit that small space of curiosity and prevent a misunderstanding that might have prevented a good relationship with the IT department.

Building a trusting relationship with the staff will take time. The consultant who invests time and effort into relationship-building will enjoy a return on investment (ROI) that consists of many intangibles like employee insights, frustrations, and awareness of existing technical problems. Owners and managers might not be aware of the employees' opinions or viewpoints about issues within the business. Never overlook or dismiss staff opinions, these represent a treasure trove of valuable information for the system design team. One gem of information from this treasure would be a picture of a problem's history, the cause, why proposed solutions failed, from a staff's perspective. Learning from the staff about past failures will prevent the consultant's team from making similar mistakes.

A few staff members will express their frustrations about issues once trust is established. The professional will not take sides, encourage, or take part in the office gossip. Consultants with selective listening skills can sort through the gossip and learn a reason for unexplained resistance to the enterprise project. Resolving issues that lead to frustration will often remove the staff's resistance toward the project and design team.

Owners and management of larger companies tend to see the organization from a strategic point of view: items like computers, networks, or how the information system works. Normally the people performing the day-to-day operations can inform you about technology issues. Employees who have had to struggle with obsolete, broken, or missing technology are eager to discuss these problems. A download about the current state of the company's technology is another gemstone from that treasure trove of staff information. Getting the opportunity to dazzle management by being able to proactively solve issues before they are problems is a great ROI for time spent building a relationship with the employees.

There is a lot of work involved in relationship-building, listening, solving problems, and addressing staff frustrations. One immediate result of this work is improved lines of communication between people, departments, management, and ownership. Frustration will create tension between employees and management impeding the flow of information or ideas within the company. Removing the points that create tension will result in a more relaxed work environment that encourages the sharing of ideas and information. The result of establishing clear communication between people and departments will be to aid in the discovery of what capabilities, products, and skills are available to the organization or clients (Schulmeisters 2014).

3.2 Capabilities

Now that people are communicating, information can flow between departments. Clear communications can help the organization or consulting team determine what is required for a capability. A capability is what an organization does as a core function. Capabilities do not define the "how" or "where"; they are concerned with the "what" is being done. Filling a truck with fuel and painting a part of it are not capabilities. These actions describe how something is being accomplished. Equipment and tool management is a capability because these tell what is being done. It is expressed in terms of outcomes and service levels to create value for either an internal or external customer.

Defining and assessing capabilities will help to discover and eliminate duplicate processes or technologies. This assessment will pinpoint gaps or weak areas in the company's skillsets and support. These capabilities provide a communication bridge between management and the staff who perform the daily business processes. Each of the company's ideas, processes, or skillsets, must be assessed to determine if it might be a capability. Once these capabilities have been identified they are ranked from highest to lowest of importance.

There are three levels of capabilities from 0 (the highest), 1 (this defines and aggregates the capabilities areas), 2 (the actual business capabilities – hiring staff, managing, etc.), 3 (this is the level where the sub-capabilities are determined) – these start to approach the level of business processes (Bedford 2014a). These business capabilities are used to create "business capability models," which is the process of modeling *What* a business does to reach its objectives (Bedford 2014a).

Each enterprise system will have its own set of capabilities that is similar to ones defined for an organization or business. Enterprise capabilities relate to the basic concern of managing the system: governance, organization, integration, compliance, validation, and finally management. Remember a capability is about the "what" of something being accomplished, not the "how" or "where" it is accomplished. Encrypting data is not an enterprise capability because it describes how data is being secured. Enterprise management is a capability since it defines what is being managed (the enterprise system).

Business or enterprise capabilities share similar traits. The traits of a capability are a definition of what, definite and specific result, a clear definition, and a unique purpose. We have already examined that capabilities deal with the "what" not "how" or "where" the action is accomplished. Another trait of each capability is to have a definite and specific result. A capability's result or outcome must be clearly defined. Equipment repaired on time will not work because there is no definite result. An example of a specific and clearly defined capability would be equipment management that will result in 85% of all machines available for work.

Every capability must be unique. Recall that the purpose of defining capabilities is to discover and remove redundant or duplication from the organization methods. Information management is very similar to data management when working with formatting, standards, and principles. It might be better to consider data architecture as an overall capability that manages both data and information assets. We will revisit the case of the Medina County Health Department (MCHD) discussed in Chapter 2 to provide examples of capabilities. Remember, the MCHD was working to establish an enterprise environment. This required establishing what are the MCHD's core capabilities. Although we are not defining the enterprise capabilities, it is a good starting point to identify what are the MCHD's organizational capabilities.

One trait of business capability is that they are unique. No two capabilities should be the same or similar. The MCHD is a public health department with a Community Health Division. The Community Health Division might have Managing Community Health as a capability statement. Unfortunately, the MCHD as the umbrella organization has identified Managing Public Health as a statement. Managing Community Health would be similar or contained within Managing Public Health. Therefore, between the two statements, Managing Community Health will be moved under Managing Public Health as a sub-capability.

Each capability must be clearly defined, and this definition is easily understandable. The capability statement Managing Public Health would be defined as "protecting the citizens against threats while encouraging healthy behavior."

All of the capabilities must have a unique purpose. Each of the business capabilities must be provided a specific definition that determines what is the capability's purpose. Any confusion about the businesses' capabilities is avoided by ensuring that there is no responsibility "overlap" or similar purpose with an existing capability. MCHD's environmental division is tasked with controlling the threat of mosquitos carrying the West Nile Virus (WNV). One capability for the division is Mosquito Control. Mosquito Control is defined as "prevent WNV outbreaks in Medina County, Ohio, by ensuring the mosquito population stays at acceptable levels." The purpose of Mosquito Control is to reduce the outbreak of WNV cases by 20% from the previous year.

Capabilities can be defined for GIS-centric businesses or enterprise systems. GIS-based capabilities must deal with the geospatial aspect of the information. This normally involves a locational relationship with the earth's surface. Consider the MCHD mosquito response program with a capability named Tracking WNV. This Tracking WNV describes what is being done. Our definition for Tracking WNV is "GIS will map county locations of WNV and the purpose is to eliminate WNV outbreaks within the Medina County."

When properly identified and defined a capability will not change. The stability of a properly defined capability allows it to become a great vision of the organization. Dynamic items like projects, processes, or EA will change as time shifts around them. Capabilities will only change if there is a transformation in the organization's mission or business model (merger or acquiring a new source of information).

Consider the case of Blue Sky Bee Supply in Ravenna, Ohio. The company was founded in 2004 by a beekeeping husband and wife team, whose careers as beekeeper suppliers began when they added friends' requested items to their own orders from beekeeping manufacturers. Blue Sky, Ltd. was an online store that shipped from the couple's garage. Due to the company's success they now operate in 2,500 sq. ft. warehouse (with showroom) in Ravenna, Ohio.

Blue Sky's basic business capability is "Beekeeping Equipment and Supplies," and it has remained the same since 2004. This capability has not changed since the company's founding in 2004. A new operational capability "Manage Warehouse" was added to the core capabilities, when they moved into the facility in Ravenna, Ohio. Even though business capabilities do not have a lot of change, it does not mean that over time the company might not add or remove them. All organizations should undergo a periodic review to ensure that they are operating with the proper set of capabilities.

Discovering and defining the organization's core capabilities is one starting point for an enterprise system build. Do not expect to discover or create many capabilities for your client. Each organization should have a maximum of around ten capabilities, more than ten and business processes are probably being mistaken as capabilities. Any organization must have a good understanding of its capabilities before attempting to build a true enterprise system.

3.3 Flexible architecture

Understanding the organization or company's core capabilities is an important piece toward creating a flexible EA. This book will make reference to flexibility as an agile system. Agility or flexibility is a critical survival trait for the system, regardless of whether it is straight-up EA or GIS. A set of well-defined business capabilities ensure that the organization is always working within its core services.

Understanding the established core business capabilities will enable the organization to quickly decide if a new opportunity can be easily exploited. When there is a perception that a large number of good opportunities have been lost or passed by, then the company should review its capabilities and determine which capabilities should be kept, developed, or replaced.

Part of the ROI for an investment in relationship-building was developing clear lines of communications within the client organization. These new lines of communication are producing results in the form of discussions about business capabilities and opportunities.

This flow of information and ideas for new capabilities and opportunities are the beginning of a flexible enterprise system. True flexibility within a system begins with good communication between people. The relationships between staff and consultants that were repaired, developed, or reinforced are paying dividends in the form of flexibility for the enterprise system. The system design team must find a way to bring all of this communication to every level of the organization's enterprise system.

The enterprise system is very similar to a living organism. When healthy the animal responds quickly to environmental stimulus for food, defense, or loss. Food for the business is successfully acquiring and exploiting new opportunities for profit. A business might need to defend its market share, products, or resist a hostile takeover. A loss that impacts profits might be a result of bad or outdated information. The enterprise system will enable the organization to control, update, acquire, and dispose of data without loss.

A well-planned EA will provide the agility that will allow the organization to succeed in a constantly changing digital world.

Agile architecture is the result of careful planning and support from all levels of the organization. Architecture strategy also impacts the agility of the architecture. Changing data into information can trigger changes among company processes. Agility is about being able to coordinate, adapt, and work with established systems (Gill 2013). The system architecture will provide a strategy that can be used to measure whether the organization should move to react or disregard the market or information change.

One important aspect of enterprise architecture is integration. A well-planned architecture provides integration and coordination among all levels of the organization, resources, and employees. A fully integrated and well-coordinated architecture will establish a strong foundation for communication among everyone involved in the enterprise. Creating and maintaining clear avenues of communication is critical to the success of an enterprise system.

A company that uses an architecture that encourages everyone to communicate concerns, ideas, or suggestions freely provides a sense of user ownership, which in turn generates trust and faith in the system (Sieber 2000). When everyone supports the system, it is in harmony, and the business receives the full benefit of all resources and personnel working together to accomplish the company's business goals. When the individuals and departments are in harmony, it becomes a system that proves "the whole being greater than the sum of parts" (Michigan Department of Information Technology 2007) and provides more significant benefits to the company.

An architecture system that is in total harmony is the result of clear and concise communication about the architecture's requirements and overall goals among everyone involved in the design and implementation process. Communication is the key to EA's success or failure. EA is about managing the company's data, risk, and response to change (Labuschagne 2011). A manager must ensure that all team members understand their roles and responsibilities on the project. When everyone has a clear understanding of their positions, the exchange of ideas and compromises can flow among the segments of the architecture plan. Management of a company's risk and response to change requires effective communication among people.

3.4 Designing an enterprise architecture

Design is the most overlooked part of implementing an enterprise system. Whether this system is pure information or EA, it must have purpose and meaning. Purpose and meaning can only come from a well-thought-out design, one that considers every aspect of the new information system and the organization's needs.

Any organization, no matter its size, will benefit from a properly designed enterprise GIS built on a foundation of EA. Achieving the support of ownership, executives, and staff is often the first and most difficult challenge a designer will overcome.

There are ways to convince stakeholders that a company should develop and implement EA. Examine daily business operations to pinpoint and locate bottlenecks of information, communication, effort, etc. that negatively impact the company. Determine where employees disagree over processes, responsibilities, fault, etc. These might be the results of miscommunication about roles, redundant policies, or duplicated procedures. Document these disagreements and the cost to solve the issue measured by time, labor, and resources. A little investigation can relate to the disagreement's root cause in the system. (An argument over a missed opportunity could have been created by both sides assuming the other handle it – confusion over responsibility.)

One should not start the investigation with the idea of finding fault. That will only lead to people becoming defensive, suspicious, and uncooperative toward solving the problem. Always consider that people will have a genuine fear of being dismissed if the owners or management perceives them as a cause of the problem. One can mention specific situations but do judge or assign blame in the final document. Use the findings of the investigation as the basis of loss analysis. Determine the total amount of time resources (equipment, goods, etc.), negative impact on the company (dissatisfied clients, the value of lost opportunities) and give each instance a value for money lost.

Demonstrate how an EA prevents this misunderstanding. Naturally, a well-designed report showing the total cost savings of an implemented system is necessary. By demonstrating how the EA can save time and resources, find new clients and opportunities, and lead to other potential revenue streams, the plan will gain the attention of investors, owners, and managers.

Once the decision has been made to implement an architecture, the enterprise architect will assemble a design team. This team might have a representative from the ownership/executive management and an enterprise, business, process, technology, and data/information architect. The team leader will be the enterprise architect who will guide the team and organization through the process of designing, implementing, and maintaining the organization's EA.

How many people it will take to fill one or more positions on this design team is determined by the size of the organization, available personnel, or the size of the available budget. It is quite possible for one person to fulfill more than one position on the architecture design team. Each position has a unique contribution to make toward the design of the EA.

3.5 Establishing goals

This section will answer the critical question of "where to begin"? Previously this chapter discussed business capabilities as a start for the enterprise project. There is a danger with starting from the capability viewpoint. When a set of capabilities has been established there is a tendency to relax and consider the design of the enterprise complete. Whereas the organization has just completed one piece of the enterprise design, not a full-system build.

A successful enterprise system build will begin with goal setting, an honest assessment of the current state, and a clear definition of people's responsibilities. Why is goal setting listed first? Well, goals provide the destination for the system, without a set of goals to work toward the system is like a ship without a compass. Goals are the compass that keeps the system's growth and development on course to the desired future vision. Goals are essential because they will set the tone for current and future state assessments, maturity modeling, and determination of final success.

One of the more difficult items about developing goals is determining what should be a goal. Potential goals can be found almost anywhere and from unexpected sources. Combing through the business's records, the business plan, documents of incorporation, quarterly and annual financial, etc. will reveal several potential goals. These internal documents are a valuable resource for identifying goals that will focus on fulfilling the company's mission and future state. Goals should help the organization improve and progress toward a shared vision of a successful future.

When working on goals, it is essential to focus on the business at a strategic level and focus on the company's capabilities. The goals should encompass the entire organization, which in turn encourages the goals to promote the company on a path to its future. When the goal search is focused only on the business units or line production, the goals are supporting a vision of the company in the current moment, which defines the current state of the company. Goals should always support the company's progress toward a future vision.

Record each goal in a document as a future reference. This list of questions defines a goal: What does it relate to? What is its name? How does it fulfill the goal? What is the expected result when achieved? How are results measured? How long to accomplish or complete? What is the source of the goal? (Bedford 2014b). Figure 3.1 is an example of how this information can set up in a goal chart. Charts or documents like Figure 3.1 provide a complete

Vision	HIERARCHY	GOAL NAME	Description of Achievement	Intended Results	Measurements	Timeframe
			CHAGRIN VALLEY ENGINEERING GIS DIVISION GOALS			
Company is stable with positive growth.	1.0	Business Stability	Eliminate wild swings between the ups and downs of a business life cycle.	steady progress forward	steady growth/ profits	5 years
	1.1	Low turn over	control turnover of clients and employees	retention of clients and employees	number of clients and employees lost	5 years
Produces a yearly profit.	2.0	Business Profitability	show increase in profits	Establish a profitable business	income over expenses	5 years
	2.1	control expenses	eliminate unnecessary expenses	increase profits by controlling expenses	increase in earnings	5 years
Enjoys steady growth.	3.0	Business Growth	Thru Clients & Products	Grow the company	how many clients & contracts per year	5 years
	3.1	Clients	Add 1 major client per quarter	Gain new clients & prevent loss	Addition of new clients each year	5 years
	3.2	Contracts	sign 2 major contracts per year	Gain Growth thru work	Number of major contracts per year	5 years
	3.3	Recurring work	Establish recurring or renewing contracts (bread & butter work)	Steady & recurring work	Number of contracts per month	5 years
Provides a range of products.	4.0	Products	applications and programs	Establish a foundation of products		5 Years
	4.1	games	mobile, desktops, online	income	1 game per year	5 Years
	4.2	applications	develop applications	income	1 application per year	5 years
	4.3	Websites	develop turnkey websites	income	1 turnkey site per month	5 years
	4.4	Mobile	Develop apps for mobile phones	income	2 applications per year	5 years
	4.5	GIS	Develop apps and programs for mapping	income	1 map app per year	5 years
Provides contract services.	5.0	Services	Consulting, IT, & GIS	Establish a range of services	Through new contracts	5 years
	5.1	GIS Consulting	Contract analysis, programming, system design	income	1 new contact every 6 months	5 years
	5.2	IT Consulting	Contract work for public & private sector	income	1 new contact every 3 months	5 years
	5.3	Information consulting	Create custom databases, information networks, data cleaning	income	3 new contacts every 6 months	5 years
A high level of satisfaction will keep customers and employees.	6.0	Satisfaction	Customer, Owner, Employee	Retention	How many repeats or employee loss	5 years
	6.1	Owner	Satisfied with company production, services, and performances.	Retention	Client & Employee comments	5 years
	6.2	Customer	Satisfied with the company's work, services, and deliverables	Retention/Recommendations	Client comments	5 years
	6.3	Employees	Satisfied with work environment, benefits, and wages	Retention/Recruiting		5 years
	6.4	Quality	Maintain the highest level possible.	Best Product Possible	Positive /Negative comments	5 years

FIGURE 3.1
Example of a goals chart.

picture of the goal. Now it is possible to determine which goals will have the most significant impact at moving the organization forward toward its desired future.

The information chart in Figure 3.1 will help sort the goals into two types: primary goals and sub-goals. Primary goals are the "big" ideas that, when accomplished, show how much progress an organization has made toward fulfilling the strategic vision. "Accomplish a 15% increase in profit" is an example of a primary goal. However, every primary goal is accomplished through many smaller steps. These incremental steps are sub-goals. The accomplishment of sub-goals will lead to the completion of a primary goal. "Eliminate unnecessary expenses and reduce accidents" are examples of two sub-goals that when accomplished, will allow a company to show an increase in profit. Primary goals and sub-goals are the navigation aids and milestones that will track a system's growth and keep it on course.

The goal statement should not be overly complicated or wordy: "Accomplish a 15% increase in profit" is too long. A goal statement or ID should be short, simple, and, ideally, no more than two words (Baer 2014) (sometimes you do have to use three or four words). "Increase profit" is much easier to say and understand statement when compared to "Accomplish a 15% increase in profit." Simple statements will prevent misunderstanding and clarify lines of communication about the expectations for each goal. The beauty of a two-worded goal statement is that they are right and straightforward to the point.

The two-worded statements provide only limited information. A goal description or definition explains the two-worded statement. The definition can be only a couple of lines or a small paragraph. If the definition requires more than five lines to explain the goal's purpose, it is time to revisit and simplify that particular goal. Clear and concise communications are our ultimate purpose of creating goals with descriptions.

Two questions that should be on your mind now is which goal is most important and how do I know? This is where goal ranking comes into play! Ranking goals is always a tough decision and it does require the input of everyone involved in planning and implementing the strategic plan. Do not be surprised if the discussion about ranking goals in order of priority generates quite a bit of passion. Discussing what is important or defining a critical goal should bring out the combativeness in people, it shows that they truly care about the organization.

One of the hardest things about goals is determining which one is the most important! Remember though, that these rankings are dynamic, will change as the system develops, and as the organization matures. So how do we rank the goals? This topic cannot and should not be answered with one person's idea or a quick note on a form. Goal rankings must be discussed and decided by representatives from the organization's stakeholders, executives, and supervisory levels. Ranking the organization's goals in order from highest to lowest priority will focus everyone's efforts on ensuring that the organization stays on course to fulfill its primary mission goal.

3.6 Case 2: The City of Oberlin, County of Lorain, Ohio

Oberlin's IT manager realized that she was losing ground in the battle to maintain the information on the city's infrastructure. The city's infrastructure information was contained in several locations, a 2006 MS Access database, AutoCAD drawings, a few GIS layers, PDF, paper: plans, notes, and even 3 × 5" index cards. New information is being accepted and added to as either an AutoCAD drawing or PDF. This infrastructure data should be available via the city engineer's office for planning, field inspections, construction projects, and the general public.

The IT manager contacted the author and Davey Resource Group, Inc "DRG" for building an enterprise GIS.

The city made an unusual decision to establish and complete a set of goals about the legacy data before beginning a full enterprise design. The goals were

1. Gather data
2. Convert into GIS
3. Load data into DRG Asset Manager software application
4. Review and edit

Once these goals have been completed, for the first time, the IT manager, engineer, and department heads will have an overview of the city's infrastructure. They will then conduct a review to identify where data is missing, edit existing data, and upload new data. Other city departments will be contacted and invited to review the information to ensure that nothing has been overlooked or that the wrong information was placed into the data archives. This review will provide the city with a starting point for creating an enterprise and data architecture.

It will be a big help for everyone involved if you use a numbering system or lettering system to show the goal hierarchy. This would be a step system where the primary goal is marked as 1, any sub-goal that relates to the primary should be assigned the label 1.1 and so on. Refer to the Goal table in Figure 3.1. Please note the column labeled GOAL ID. This column displays not only the goal ID but how the goals relate to one another. When you use a program like VISO, the numbers in the ID column will create a wonderful visual graphic of the relationship between goals.

It is important that everyone can see, visualize, and understand how the goals relate to one another. Information contained in the chart will help determine when a goal has been achieved. This information will help to determine if achieving the goal produces the results planned or if there was an unexpected impact on the company. These are not set in stone, if the progress or results are not expected or desired then goal should be changed or removed. New goals need to replace the ones that are old or completed. The goal management is never ending; however, it is vital to the success of the enterprise system and the business.

3.7 Summary

This chapter defined EA and explained how it can benefit an organization. Readers learned how an agile and adaptable system can provide a market advantage in today's digital world. A small discussion was presented that provided a few of the reasons why companies or organizations chose to implement EA. The chapter also discussed establishing goals and the beginning step to develop an EA. Finally, the author pointed out that the design stage is very important yet often the step many skip in the rush to build the enterprise system.

3.8 Assignment

Assignment 3: The company

This assignment will be the first step on the road to designing and implementing an EA for the enterprise GIS. The best way to learn about enterprise architecture is to create one! This assignment will have several parts. When you have completed each part you will have a company to use in future assignments.

Think about or create a business, non-profit, or another type of organization that you might like to own, work for, or know well. Give your organization a name, and then write a one or two paragraph describing the company (what is it, what are the products, where is it located, how many departments, etc.)

Assignment 3.1 Establishing goals

Create five primary goals with two sub-goals for this organization.

Bibliography

Baer, D. 2014. *Enterprise Architecture EACOE*. https://architecturescoe.org/.
Bedford, D.P.D. 2014a. *Business Architecture*. Kent, OH.

Bedford, D.P.D. 2014b. *Business On a Page (BOAP)*. Power Point. Kent State University, Kent, OH.

Dueker, Kenneth J., J. Allison Butler. 1997. *GIS-T Enterprise Data Model with Suggested Implementation Choices*. Portland, OR: Center for Urban Studies, School of Urban and Public Affairs, Portland State University, October 1.

Gartner, Inc. 2008. "Gartner Clarifies the Definition of the Term 'Enterprise Architecture'." *Gartner Research* (Gartner), 15. doi:G00156559.

Gill, Asif. 2013. "Defining a Facility Architecture within the Agile Enterprise Architecture Context." *Orbus Software*. October 1. doi:WP0107.

Labuschagne, Louw. 2011. "Building Enterprise Architectures for Non-Architects." *Orbus Software*. September 1. doi:WP0011.

Lapkin, Anne, Philip Allega, Brian Burke, Betsy Burton, R. Scott Bittler, Robert A. Handler, Greta James. 2008. "Gartner Clarifies the Definition of the Term 'Enterprise Architect'." *Gartner*, August 12: 1–5.

Michigan Department of Information Technology. 2007. *From Vision to Action: Enterprise Architecture – Strategic Approach*. Michigan Department of Information Technology.

Rouse, Margaret. 2007. "Enterprise Architecture (EA) Definition." *TechTarget*. June 01. Accessed October 2015. http://searchcio.techtarget.com.

Schulmeisters, Kar. 2014. "Enterprise Architecture: Bridging Entrepreneurs and Hard Problems." *Orbus Software*. November 1. doi:WP0168.

Sieber, R.E. 2000. "GIS Implementation in the Grassroots." *URSIA Journal* (Vol. 12, No. 1): 15–29.

The Open Group. 1995–2015. *Enterprise*. January 1. www.opengroup.org/subjectareas/enterprise.

Thomas, Christopher. 2009. "What's Your Definition? Looking at What Enterprise GIS Really Means." *ArcUser*, 3032. http://www.esri.com.

Wilson, N. 2013. A GIS manager's dialogue: Is your business semi-integrated or fully system integrated? *North Carolina GIS Conference*. Raleigh, NC.

4

Roles in Enterprise Architecture

4.1 Roles and responsibilities

Chapter 4 will introduce and discuss the roles and responsibilities of each person participating in enterprise architecture (EA). This chapter will discuss and demonstrate how ownership and executive management support will determine the success or failure of the EA development project. It defines the primary roles of the enterprise architect, business architect, and data architect and discusses how they must cooperate for the EA to succeed is. This section of this chapter will demonstrate how vital establishing clear communication between management, supervisors, and workers will be for the EA to be successful. The chapter makes the point that excellent communication only exists when everyone understands his or her roles and responsibilities in a system.

Once a vision of the organization is defined, it is time to determine who is responsible for each part of the concept (Sieber 2000). Establishing and defining roles and responsibilities is vital for creating the architecture's structure and chain of command. People need to know and understand how the flow of authority and ideas will move throughout the organization. Communication will also enable accountability; no one can claim that they were not told or did not understand their role. Their roles and responsibilities are laid out for everyone, and if there is a problem living up to expectations, everyone can see the failure. Determining roles and responsibilities is defining who will be the organization's leaders who bring architecture to life.

The only way to develop enterprise architecture and bring it to life is by filling the positions within the architecture. Each position carries with it a certain amount of responsibility, and the person who assumes the role must complete the associated tasks. Placing staff into positions of trust is also a significant aid in achieving the ultimate "buy-in" and support from the organization's rank and file. Areas of responsibility provide "staff/low-level employees" to shine and work with the executive level, thereby creating more precise lines of communication among all levels of the organization.

Consider the previously mentioned process architect position. The person assigned to fill this role will need a working knowledge of the daily operations and be able to develop a good working relationship with other staff members. A knowledgeable staff person would be able to map out the day-to-day business processes and enhance this process knowledge by meeting other departments. Figure 4.1 is an example of a roles and responsibilities chart (Bedford 2014).

The chart is an example of how elements within the enterprise architecture can be tied together. The top role shows what question is being asked and answered when designing the architecture. Just under the top row with italic headings are parts of the architecture strategy. The first column contains a description that defines a goal or capability. The last section of the Figure 4.1 is for the role or position that is responsible for architecture planning.

You are working to create data models for a new architecture when someone asks a question about what type of computer processor or hard drive should be purchased. Since you don't know that answer, you can check Figure 4.1. A quick look down the left column brings you to "Identify Technology – Tech Specific," now travel to the left, and you will learn that the technology architect is responsible for answering the questions of how and why a specific processor or hard drive is required. This architect would also be liable to explain how the computer and hard drive will impact the architecture's "strategies and goals" while enhancing the "processes and activities." Traveling across the chart will reveal the other architects and their areas of responsibility when a new piece of technology is purchased. Roles and responsibilities tables can have different formats, but they all help to answer the ultimate question, "Who is responsible for what?"

Now, this author understands that the readers are asking themselves what does this have to do with developing or implementing an enterprise GIS? The answer is quite simple EVERYTHING! When designing an enterprise GIS, the key positions (roles) are defined, and their corresponding responsibilities assigned. Defining the responsibilities for each key position provides a checklist of skills that the person in the position should have. A clear definition of responsibility ensures people in crucial roles accountable for their actions and decisions. Listing the responsibilities also removes the convenient excuse "I did not understand that was my area of responsibility" and protects the organization from people who "polish" the resume when pursuing jobs. Defining responsibility is a way to ensure that people fulfilling the key roles can be held accountable to everyone within the system.

How does the idea of roles and responsibilities transfer over to the world of GIS? One begins by listing every possible position for a fully functional GIS department (Table 4.1). The chart in Table 4.1 shows the title and description or responsibilities of each position connected to the GIS division. Remember, one of the advantages of designing architecture or a strategic plan is scalability! A small department of one or two can use this chart to determine growth while a more significant division can use this chart to clear lines of communication and eliminate areas of responsibility or authority overlap.

The next step is to create a chart that replaces the column about Alternate Titles with EA responsibilities or descriptions. Refer to Table 4.2 as an

Roles in Enterprise Architecture

ARCHITECTURE RESPONSIBILITY CHART	WHY STRATEGIES & GOALS	HOW PROCESSES & ACTIVITIES	WHAT MATERIALS & THINGS	WHO ROLES & RESPONSIBILITIES	WHERE LOCATIONS & GEOGRAPHY	WHEN EVENTS & TRIGGERS
DESCRIBE - THE BUSINESS	ENTERPRISE ARCHITECT	ENTERPRISE ARCHITECT	ENTERPRISE ARCHITECT	ENTERPRISE ARCHITECT	ENTERPRISE ARCHITECT	ENTERPRISE ARCHITECT
DEFINE - THE RELATIONSHIPS	BUSINESS ARCHITECT	BUSINESS ARCHITECT	BUSINESS ARCHITECT	BUSINESS ARCHITECT	ENTERPRISE ARCHITECT	ENTERPRISE ARCHITECT
SPECIFY - COMPONENTS & SERVICES (TECH-NUETRAL)	BUSINESS ARCHITECT	BUSINESS ARCHITECT	DATA/INFO ARCHITECT	APPLICATIONS ARCHITECT	BUSINESS ARCHITECT	BUSINESS ARCHITECT
IDENTIFY - TECHNOLOGY (TECH SPECIFIC)	TECHNOLOGY ARCHITECT	TECHNOLOGY ARCHITECT	APPLICATIONS ARCHITECT	APPLICATIONS ARCHITECT	DATA/INFO ARCHITECT	BUSINESS ARCHITECT
SELECT - SOLUTIONS	BUSINESS ARCHITECT	TECHNOLOGY ARCHITECT	DATA/INFO ARCHITECT	APPLICATIONS ARCHITECT	TECHNOLOGY ARCHITECT	BUSINESS ARCHITECT
SOLUTIONS	SOLUTIONS	PROCESSES	DATA	WORK PRODUCTS	TECHNOLOGY	EVENTS
THE ENTERPRISE	GOALS	PROCESSES	MATERIALS	ROLES	LOCATIONS	EVENTS

FIGURE 4.1
Roles and responsibilities (Bedford 2014).

TABLE 4.1

GIS Department Roles and Responsibilities

Roles are used in models to show who performs the task associated with a capability or process. A role normal identifies a single staff position that performs the operation; however, a role may also describe an operation that is performed by more than one staff position.

Role	Description	Alternate Titles
GIS Manager	Description of Roles: Oversees and manages all aspects of the GIS department operations, is the liaison between unit staff, engineers, surveyors, CAD techs, and company owners.	Varied: GIS Coordinator, Department Head
GIS Analyst	Description of Roles: Assists GIS manager in supervising GIS operations and staff and project manager for GIS projects, performs advanced GIS analysis and mapping, recommends staff to project vacancies, oversees training for GIS staff and interns.	Varied: Assistant Manager, Senior Staff, Senior Lead, Team Leader
GIS Technician	Description of Roles: Performs GIS analysis and mapping, manages interns, manages data collection procedures.	Varied: GIS Analyst 1, Senior GIS Tech, Assistant Team Leader
GIS Data Steward	Description of Roles: Safeguards datasets from errors, ensures data is accurate and up to date.	Varied: Wetlands Steward, Engineering Steward, Project Steward, Data Liaison
GIS Applications/ Tech Manager	Description of Roles: Maintains application and technology inventory, ensures lifecycle is followed, implements changes, troubleshoots issues.	Varied: IT Architect Applications Tech, Tech Coordinator, Applications Manager, IT Manager

example of a combined roles table. This table will be the first step in merging the roles required for developing the enterprise architecture with the established positions within a GIS department. It is crucial that the EA's design proceeds. The enterprise GIS architect must also determine the roles that will be needed to complete the system architecture and the position's responsibilities. He or she can use the format that was introduced in Table 4.1 as mentioned earlier or create their template like Table 4.2. The architect will meet with the executive management and ownership to discuss what positions are needed, responsibilities, and whether they should be filled internally or through new hires. It does not matter how one formats the table, chart, or summary as long as the organizational roles and responsibilities are clear.

Establishing roles, defining and assigning responsibilities, provides a method that not only communicates expectations about people who are holding positions of authority, but also creates accountability for mistakes, delays, or sub-par performances. Defining and assigning roles at the beginning of the enterprise system design process establishes a chain of command and communication. A chain of command is required to ensure clear lines of communication, exchange of ideas, and ensuring that projects meet

TABLE 4.2

Combined GIS and EA Roles and Responsibilities

Roles are used in models to show who performs the task associated with a capability or process. A role normally identifies a single staff position that performs the operation; however, a role may also describe an operation that is performed by more than one staff position.

Role	Potential EA Role	EA/GIS Roles Description
GIS Manager	Enterprise Architect	Description of Roles: Oversees and manages all stormwater of the GIS department operations. This person will oversee the design and implementation of the strategic plan for data and information.
GIS Analyst	Business Architect/ Data Architect Subject Matter Expert	Description of Roles: Assists GIS manager in supervising GIS operations and staff and project manager for GIS projects, performs advanced GIS analysis and mapping, recommends staff to project vacancies, oversees training for GIS staff and interns
GIS Technician	Data Architect/ Applications Architect/ Data Steward/Subject Matter Expert	Description of Roles: Performs GIS analysis and mapping, manages interns, manages data collection procedures.
GIS Applications/ Tech Manager	IT Architect/Applications Architect/Subject Matter Expert	Description of Roles: Maintains application and technology inventory, ensures lifecycle is followed, implements changes, troubleshoots issues.

deadlines. Staff members need a command structure or framework to understand how they fit into the plan and how their efforts will help the system. A robust organizational hierarchy will keep confusion, miscommunication, and misdirection from crippling the enterprise system.

The author would like to make the point that not all of the positions necessarily are filled by a staff member. It is possible that in smaller organizations one person might have to fulfill the role of two or more of these positions. This is not due to a lack of desire to build out the GIS into a full-enterprise GIS; it is simply a matter of there not being enough money to hire more staff. One example of this situation is found in the City of Oberlin. The city's IT manager is also the GIS coordinator, and AutoCAD technician is responsible for all of the drawings to support construction projects within the city.

When the city decided to implement an enterprise GIS project, the organizational chart for the project roles contained positions filled by contractor staff and other city departments.

The organizational chart in Table 4.3 reveals the challenges of establishing and implementing an enterprise project with a department of one person. The consulting team can step in and fill those positions during the build phase and while the company is under contract with the city. The subject matter experts (SMEs) are from the water department and public works

TABLE 4.3

City of Oberlin Enterprise GIS Organizational GIS Roles

Roles are used in models to show who performs the task associated with a capability or process. A role normal identifies a single staff position that performs the operation; however, a role may also describe an operation that is performed by more than one staff position.

Role	Description	Employer
IT/GIS Manager	Description of Roles: Oversees and manages all aspects of the GIS department operations; is the liaison between unit staff, engineers, surveyors, CAD techs, and company owners.	City of Oberlin Public Works Department
Enterprise Architect	Description of Roles: Oversees the design and implementation of the City's Enterprise GIS.	Contractor
GIS Analyst	Description of Roles: Performs GIS analysis and mapping, manages interns, manages data collection procedures.	Contactor
Subject Matter Expert	Description of Roles: Safeguards datasets from errors, ensures data is accurate and up to date.	City of Oberlin Public Works
Subject Matter Expert	Description of Roles: Safeguards datasets from errors, ensures data is accurate and up to date.	City of Oberlin Water Department
GIS Applications/ Tech Manager	Description of Roles: Maintains application and technology inventory, ensures lifecycle is followed, implements changes, troubleshoots issues.	Contractor

(storm water coordinator). This is the challenge faced by many small GIS departments finding people to fill the positions and fulfill those additional responsibilities that are required to establish an enterprise GIS.

The contractor's staff will ensure the city's enterprise GIS project is implemented and working properly. The city then has to find a way to ensure that the enterprise GIS is maintained properly. The city would then have to pay for the consultant's design team to return and fix anything that was broken during the mayor's speech.

4.2 Building the team

"There is no I in the word team!" – coaches, managers, and consultants often quote this old expression, and many others tasked to create a team. This expression is very true in a literal sense, for letters T-E-A-M make up the word. There is no "I"; however, the real message that has resounded across our sports culture for generations is that one person alone cannot accomplish the goal (or winning); it takes a team (group) of dedicated people to achieve final victory. One must always consider this fact about teams: they are created by persuading a group of people to work together. Each member of a team is a unique individual with his or her own beliefs, strengths, weaknesses, opinions, and possibly their personal goals. Successful managers, coaches, and leaders understand this point and provide leadership that accounts for each team member's uniqueness while ensuring everyone contributes to the team's success. So now is the time to discuss each position's potential GIS and enterprise architecture responsibilities.

4.3 GIS manager/enterprise architect

Let us start at the top of the organizational chart with a position that will lead and provide guidance to the enterprise system's development team. This position will require someone who can multitask! The development team's leader must be able to manage the GIS department while developing a strategic plan (enterprise architecture) for data and information flows throughout the department or organization. Leading the development of an enterprise system and a department is not an impossible requirement. However, it does require someone who can split their attention from a tactical (daily operations) to a strategic (entire organization) perspective. However, they must be a leader who will find a way to merge a set of unique personalities into a cooperative or team effort.

These qualities in a team leader cannot be shortchanged or overlooked; whoever is chosen (or given) the role of the GIS manager/enterprise architect (GISM-EA) will not only be responsible for developing the enterprise system plan, overseeing the plan's implementation, but he or she will also be responsible for the selection and development of a team to build the system. A GISM-EA must have a few people skills like listening, understanding, empathy, trust, etc. and, most importantly, the ability to evaluate talent. These responsibilities manager of the enterprise design project is also responsible for communicating with each partner to ensure that new enterprise architecture meets their expectations while ensuring it supports the company's needs, goals, and daily operations. The GIS/enterprise architect will meet and build a working relationship with the company's process technology, data architects, and department heads (Ross et al. 2006). The enterprise architect will work to ensure that the architecture plan addresses their concerns.

The following section will discuss the members of a GIS department who assume the responsibilities of an enterprise architecture role. Like a ship's captain, an enterprise architect is responsible for developing the overall strategy and strategic vision of the new enterprise system. He or she is responsible for ensuring that the development strategy performs to the executive management's expectations. Let's leverage the great Star Trek debate about who is the better Captain Kirk or Picard. Captain Kirk is The Original Series (TOS) commander of the USS Enterprise NCC-1701. He often disregarded orders or the directions from higher authority orders if he felt that his actions were correct. Captain Kirk often used the defense that the "end justifies the means."

Captain Picard from Star Trek Next Generation (STNG) had a more strategic view of the world. When he needed to make a decision, all of his command officers gave an opinion, and their ideas influenced his decision. Picard would then make a final decision based upon which answer best met the strategic objectives of the situation.

The question is which approach do you feel more comfortable as being the guiding source for the enterprise system. Do you think that a GIS manager who is an impulsive, yet brilliant, tactician in the style of Captain Kirk properly guide the efforts to create and implement an enterprise system? What about the option of a manager who adopts the technique of Captain Picard? Picard represents a more reserved and strategic personality. He will consider all the options and follow the choice that meets the overall mission objective. A great leader is also a great communicator! Anyone who aspires to become a leader must be willing to listen as well as direct.

4.4 Business architect

The business architect will handle weekly meetings with the process technology, data/IT architects, and department supervisors. Weekly meetings will allow him to monitor whether their divisions are complying with the architectural standards and principles. These meetings will also allow each architect or department supervisor to "clear the air" about any issues or concerns they see with the architecture plan (Bedford 2014). The business architect will bring unresolved matters to the attention of the enterprise architect.

The business architect will assemble a team to design and implement a business architect plan that supports and works toward the goals laid out in the enterprise architecture. The business architect will also oversee the process for obtaining a compliance waiver for exceptions to complying with the enterprise architecture. The Architecture Review Board will supervise the waiver process and report to the business architect (Bedford 2014). The business architect will have to balance time between fulfilling his or her responsibilities and being the enterprise architect's executive office.

The business architect is often considered the right man/person of the enterprise architect. He or she performs many of the functions of a ship's executive officer (XO). The various department heads report directly to the XO; it is the XO's responsibility to ensure that there are clear lines of communications between the enterprise architect and each of the department heads. A good XO finds solutions to a problem before the situation requires the attention of the captain.

Let's consider the two primary executive officers in the TOS or STNG, Commander Riker and Spock. Spock is known for his logical and scientific approach to any situation, whereas Riker is considered a younger version of Captain Kirk, wild, quick-tempered, and quick to act, often willing to bend a few rules. Spock and Riker, despite being very different in personalities and approaches to life, are considered excellent executive officers with full expectations of becoming future captains.

A business architect within an enterprise system must encourage communication between people, department heads, and the enterprise architect while ensuring that smaller problems are resolved before reaching the enterprise architect. What is the best option for someone to be selected as an XO or business architect?

4.5 Data architect

The data architect will establish standards and principles for data/information throughout the company. These protocols will cover data cleaning (scrubbing), use, distribution, replacement, and security. He will ensure that his data protocols support and comply with the strategic goals and vision of the enterprise architecture. This person will also build the models that will determine how data will flow into the system and becomes information.

The data architect is not a position that should be taken lightly nor awarded as a favor. A data architect will hold the future success or failure of the enterprise GIS in their hands. Data architecture requires patience, attention to detail, and the ability to listen and combine multiple viewpoints into one cohesive system model. A good data architect must also be able to step away from the technical jargon and "simplify" the language so anyone can understand how it works.

The question is who has the skills of a data architect, yet can sit down and talk with the "blue collar" workers who will be using the system. Once again, this becomes a personality question because the enterprise architect must find someone who is technically skilled yet able to be a mentor to people without a technical or GIS background. How to find a particular person with this skillset is the challenge faced by the enterprise architect and human resource department.

4.6 Data steward

More than one person can fill the role of a data steward. A data steward will be the person responsible for maintaining accuracy and completeness, and ensuring that everything is up to date. A steward should be determined for each section or department (i.e., wetlands, engineering, city service department, etc.). An extensive enterprise system receiving large amounts of data from several different sources could quickly overwhelm a single data steward.

The position of data steward can be beneficial to an enterprise architect designing, building, and implementing a new enterprise system. It is not uncommon in local governments or private corporations that one person, who due to their length of tenure, unique skillset, or attitude, claims ownership of a particular dataset. Unfortunately, these established employees are often resistant to a new system or change. These are the people an enterprise architect must convince to support the new system. When someone is resistant to change or a new system, it is because they feel threatened.

A person resistant to the new system or change might be afraid of losing their position of authority within the organization. It is not an uncommon situation when implementing a new system that promises to streamline organizational processes. This perception is the result of designers dehumanizing employees or staff by referring to them as corporate resources. The team understands that when a system designer decides to eliminate redundant "resources," they are eliminating jobs and people. The tactic of calling employees resources helps accountants, managers, or company administrators feel better about reducing staff; it does not help an enterprise system receive support or "buy-in" from the team.

Place yourself in the position of these employees. Several of your fellow staff members can perform your role in the company. People do not trust a new system that promises to eliminate redundant functions and duplication of effort, through a streamlining of the work process. A "streamlining" implies that you or fellow workers will be the redundant or duplicated "resource" eliminated. Overlooking or forgetting that a new system impacts people, not resources, is a cardinal sin, committed by too many designers.

A little anticipation and nimble footwork can avoid this situation. The trick is to turn a strong opponent into an even stronger ally! How may you ask? By merely recognizing their position, unique skillset, tenure, etc. By offering them the role of a data steward, allowing you to quietly recruit a "watchdog" over critical data while soothing ruffled feathers through recognizing their experience, skills, or unique position. A little recognition or acknowledgment can go a long way in gaining the support of difficult people.

4.7 Subject matter expert

A subject matter expert or SME is a person who has a deep and thorough knowledge of a particular subject, field, area, or skill. Good SMEs live and breathe his or her specialty, and will work hard to keep their knowledge of the subject updated. They are invaluable assets because they can be counted upon to provide informed opinions, options, or solutions to problems encountered by an organization. Designating someone as an SME is also an excellent way to promote employee support or "buy-in" to a new system.

The problem with SMEs is determining when a person is an actual expert and not a run of the mill "know it all"! A real expert can discuss his knowledge without having to self-promote, degrade opposing viewpoints, or always use vague "I read it somewhere" sources. SMEs might not be able to recall a reference in a second, but will be able to produce a correct answer without relying upon secondary or vague sources. They are not offended when asked to provide source material and will readily share information sources.

There is not a shortage of people who will volunteer to be SMEs, and there is a shortage of people who are qualified to fulfill the role of SMEs. The trick is to determine the difference when deciding who will be an SME in your system. The process of finding a good SME is very similar to the hiring process to fulfill an open staff position. One should begin with a review of the SME candidate's resume, paying particular attention to the amount of direct experience that is relevant in the field. Remember, a resume is a summary of their work and related field experience.

A resume is not enough to determine if someone is qualified to become an SME. A deeper dive into their interests and off-work activities is required. Do they participate in the field on their own time? Have they published, blogged, posted, present at conferences or seminars, teach, etc. about the field? What is their reputation in the field? Are they considered knowledgeable persons by peers? These are essential questions that must have the right answers before deciding a candidate can be declared a qualified SME.

4.8 Applications/technology architect

A GIS application/technology architect will be responsible for the applications and technology catalog. The applications catalog is a listing of all the apps that are used by or interact with the enterprise GIS. This inventory assigns each software application a unique identification number and determines how it impacts the GIS, how many licenses are available, who uses it, what other applications need it, its reliability, and when it can be considered obsolete (end of lifecycle). When completed, this catalog will display all of the software applications used within the enterprise system and double as a budgetary planning aid by revealing when applications need to be upgraded or replaced.

This position can be split into an application or a technology architect. Some designers consider these positions as subject matter experts. Although the applications/technology architect is mainly responsible for ensuring that the software applications and technology are either up to date or capable of supporting the organization's information needs. The best people for this position "live and breathe" the deep inroads of software applications and IT technology. These are the people that everyone goes to for the "inside information" on the best application, technology, or method to set up home networks or the best app for home movies.

System designers must resist the temptation to simply "slot or award" friends and co-workers positions on the enterprise team. Although working with friends or co-workers might make things more comfortable in the beginning and short term, over time, this shortcut will work against system success. Friends or people who are "awarded" these enterprise design positions might not be willing to express their disagreements, different viewpoints, or ideas fully because they are afraid of offending their "benefactor" or losing the job. These people tend to be followers as well since they might not have the proper background to fulfill the responsibilities of the position. This tendency to follow or find the most "pleasing" answer will become a detriment to the design process.

New ideas are not born in a calm, agreeable, or passive environment! They are born from the cauldron of disagreements and passionate defense of a point of view. These conversations can be loud and boisterous, but they should never descend into personal attacks or disrespect. In-depth discussions are where the polished communication skills of a mediator or facilitator (enterprise or business architect) come into play. He or she must allow the debate to develop while ensuring everyone's idea is heard and given equal consideration.

4.9 Owner and executive management

One big challenge facing anyone desiring a new system or implementing wholesale changes will be gaining the support of the owner(s) and executive management. Owners, chief executive officers (CEOs), chief financial officers (CFOs), etc. are concerned with many strategic issues and, of course, how much profit or loss incurred. Usually, several layers of management insulate these executives from the rest of the organization. Face it, owners and top-level executives are at the top of a company food chain.

A secondary executive level might exist at the company. This level would be made up of the junior partners or assistants to the executives. These are the people who often advise or summarize new ideas for the executive level. A potential ally for a system designer would be the administrative assistants, sometimes known as executive secretaries are unspoken members of the executive level. Executive assistants with a long tenure in their positions quite often will have some influence on how "their" executive might view a new idea or system. Administrative assistants will often perform the duties of data entry, formatting, or quality control. Convincing or demonstrating to these assistants that a new system will make their jobs easier can help to develop system buy-in from the top levels of company management.

Then there are regional managers, store managers, and department supervisors. The last level of management includes field supervisors, team leads, and special coordinators. These are people on the first rung of the leadership ladder. Together these management layers make up the brain trust of the company and will ultimately decide the fate of the enterprise system.

The owner will be the highest level of the company and the enterprise architecture. They will determine if the strategic vision of the architecture plan is in line with their expectations of the company's future. Smooth operation and implementation of the enterprise architecture occur if each of the owners demonstrates their support of the architecture plan. These partners will bear the responsibility of ensuring the staff's acceptance and compliance with the architect's governance process, standards, and principles.

4.10 Department supervisors

Each department supervisor will work to ensure that the departments support and contribute to the enterprise architecture's strategic visions and goals. They will provide the local control and quality checks needed to ensure the organization's daily operations are supported by and support the enterprise architect.

4.11 Summary

The management and the business architecture team will need to work together to oversee the implementation of architecture policies and procedures. These policies and procedures will ensure that company policy, employees, services, and products remain compliant with enterprise architecture ideals. This oversight will help provide insight for the enterprise architect when adjusting, updating, or improving the architecture's present and future states.

4.12 Assignment

Assignment 4: Why

The class breaks into teams that consist of three students. Each group should read the following paragraph.

Scenario:

You and your team members have been working at an organization for several years. Co-workers have expressed frustration about duplicated efforts, lost paperwork, inaccurate data, and information that is of questionable value. There are daily arguments within the company over who is responsible for what and why every project seems to be behind schedule.

Each team should write a one-page paper that explains to management (owners, executives, and supervisors) how an enterprise architecture would answer the problems in the scenario description.

Assignment 4.1: Roles

Go back to the company you created or decided to use in assignment 3. Defining the roles (positions) and their associated responsibilities is the best way to prevent confusion among staff about areas of responsibility.

Create a chart with the following columns: role/position, role description/responsibility, an alternate title. (You have the option of breaking the role description/responsibility into separate columns.) Consider all of the roles/positions that are available in the company or a department. Then fill out your chart with the information.

Bibliography

Bedford, Denise. 2014. *Business Architecture*. Kent, OH: Bedford, Denise Ph. D.
Dueker, Kenneth J., J. Allison Butler. 1997. *GIS-T Enterprise Data Model with Suggested Implementation Choices*. Portland, OR: Center for Urban Studies, School of Urban and Public Affairs, Portland State University, October 1.
Gartner, Inc. 2008. "Gartner Clarifies the Definition of the Term 'Enterprise Architecture'." *Gartner Research* (Gartner) 15. doi:G00156559.
Gill, Asif. 2013. "Defining a Facility Architecture within the Agile Enterprise Architecture Context." *Orbus Software*. October 1. doi:WP0107.
Labuschagne, Louw. 2011. "Building Enterprise Architectures for Non-Architects." *Orbus Software*. September 1. doi:WP0011.

Lapkin, Anne, Philip Allega, Brian Burke, Betsy Burton, R. Scott Bittler, Robert A. Handler, Greta James. 2008. "Gartner Clarifies the Definition of the Term 'Enterprise Architect'." *Gartner*, August 12: 1–5.

Michigan Department of Information Technology. 2007. *From Vision to Action: Enterprise Architecture – Strategic Approach*. Michigan Department of Information Technology.

Ross, Jeanne W., Peter Weill, David Robertson. 2006. *Enterprise Architecture as Strategy: Creating A Foundation for Business Execution*. Boston, MA: Harvard Business School Press.

Rouse, Margaret. 2007. "Enterprise Architecture (EA) Definition." *TechTarget*. June 01. Accessed October 2015. http://searchcio.techtarget.com.

Schulmeisters, Kar. 2014. "Enterprise Architecture: Bridging Entrepreneurs and Hard Problems." *Orbus Software*. November 1. doi:WP0168.

Sieber, R.E. 2000. "GIS Implementation in the Grassroots." *URSIA Journal* Vol. 12, No. 1: 15–29.

The Open Group. 1995–2015. *Enterprise*. January 1. www.opengroup.org/subjectareas/enterprise.

Thomas, Christopher. 2009. "What's Your Definition? Looking at What Enterprise GIS Really Means." *ArcUser*, 3032. http://www.esri.com.

Wilson, North Carolina. 2013. A GIS manager's dialogue: Is your business semi-integrated or fully system integrated? *North Carolina GIS Conference*, Raleigh, NC.

5
Data Architecture

5.1 Data architecture

Chapter 5 will continue the process of building an enterprise architecture (EA) within the enterprise GIS. This chapter's beginning will open by defining the difference between data and information. It will then move into explaining and demonstrating the importance of how adequately applied data formats, standards, and principles, transform useful data into valuable information. This chapter will illustrate how these data procedures fit into the enterprise GIS's geodatabases and practices.

This chapter will look at the difference between data and information. How does one transform data into information? There are different types of data within an information system. Data architecture plays a significant role in managing, organizing, and converting data into information.

Data architecture is a sub-architecture that works inside of the EA to manage the collection, creation, and storage of data within an organization (Bedford 2014a). The management of data is the primary concern of any data architecture. The data architect system model will also provide a model that guides the process of changing organizational data into organizational data.

The data architect is responsible for all of the standards and principles for data/information in the company. These protocols will cover data cleaning (scrubbing) use, distribution, replacement, and security. He will ensure that his data protocols support and comply with the strategic goals and vision of the EA.

Data and information are two significant resources in the overall EA. Data converted into accurate and reliable information will impact almost every decision made at each level of the organization. The crux of the data and information is how to keep things organized, adequately captured, easily accessible to authorized people, accurate, and up to date. Data architecture will enhance the entire EA.

Converting data into valuable information that can help the organization adapt and thrive is the heartbeat of the data architecture. Reliable, accurate, and current information is the result of a good set of data. Good clean datasets are the result of establishing data standards, principles, and practices.

The data architecture will develop data standards, principles, methods, and governance policy.

Data architecture is a sub-architecture found within the EA (Bedford 2014b). It is the blueprint or strategic plan that coordinates all of the data resources within the organization. This plan defines the standards, principles, procedures, formatting, and data flow throughout the organization. The data architect is responsible for creating and implementing the data architecture, and he or she reports directly to the business or enterprise architect.

5.2 Data vs. information

This is probably the best place to discuss data and information. It is essential to understand the difference between data and information. Data is perhaps the most misunderstood or misused word, thanks to the TV shows Star Trek (also known as The Original Series [TOS]) and Star Trek The Next Generation (STNG). TOS featured the character Spock who was a very logical person and a master of "data," while STNG featured the character data, an android who strove to understand what it meant to be human. Spock and Lt. Cmdr. Data is a master of solving complex problems by staying calm, using infallible logic, and knowing the data.

Time and time again, Spock and Data saved the ship and their captains' careers with a masterful stroke of data interpretation. Unfortunately, in the world outside of TV or Hollywood, data does not provide the answers that save ships or solve cliff hangers! Data on its own has no meaning or substance. Data must be given context to have meaning. The number nine has no meaning; it could be describing pigeons, pennies, people, etc. The data point, called nine, receives context by combining it with other pieces of data. When data is interpreted and provided context, then it is transformed into information.

5.3 Enterprise data

Data within any organization is not confined or found in one specific level, quite often a piece of data in engineering, might be helping solve a problem in environmental science. When a part of data has value for the entire organization, it has become enterprise data (Graham 2012). Enterprise data can be given context by anyone or department to create departmental level information, which can influence organizational business decisions.

Every piece of data can influence for good or bad the organization's information. When one part of bad data can corrupt the entire information dataset, it is a priority to protect against data errors. Preventing data corruption is the reason for implementing a data architecture that defines how data can be collected, stored, or used. The architect, with his data plan (architecture), is the first line of defense for an organization's critical data.

The last section introduced the concept that a piece of information (data) could exist at the enterprise level (Graham 2012). The digital age has transformed companies' networks from stand-alone departments (silos) into large, complicated, and interconnected systems of information that can stretch from corporate headquarters to locations anywhere on the globe. Today's data exists at many different levels in the modern information network.

We have alluded to the fact that data can exist at different levels within an enterprise or information system. Sharing large amounts of data is one of the effects of today's modern information networks. The archaeologist measurements of an ancient roadway might become part of the information base used by transportation engineers determining the best route through a swamp, forest, mountain range, etc. The point is that today's data can be gathered and used by anyone from anywhere!

The biggest question to be answered is what level or type of data are they gathering and or using in their business decisions. The book "The Enterprise Data Model" by Andy Graham is an excellent resource for a short and concise explanation of what types of data are at certain levels in an organization. Graham outlines six different data types for each level of an organization. Breaking the data down by these categories will provide greater control and flexibility to system governance.

Remember, this book is about building an EA as a foundation for an enterprise GIS. When a GIS becomes an enterprise system, it will be transferring, accepting, processing, and archiving data for every level of the organization. The ultimate goal of this enterprise system is not only to collect and store data but to provide context, thereby creating a system of information.

The fact that a system reaches every level or department within an organization does not mean that it provides enterprise data. The critical point to remember is that data is only considered enterprise data when it provides something of value to everyone within the organization (Graham 2012). Consider your local municipality that maintains a current record of its infrastructure. The city has maintained a project history within the GIS of every road repair/upgrade project performed on each city street. This information can be used by the city's engineer to plan and budget the next year's maintenance projects. The finance department can use the accounting records to determine a baseline of how much money should be budgeted and how much contractors should be charging for materials. The service or street department can use the information to determine how much human resources and money to budget for street maintenance and striping projects. These projects are a small sample of the departments that would receive a value from the project history data for the streets. Data becomes an enterprise dataset if more than the ownership level of an organization can find benefits from using the data.

5.4 Metadata

Any enterprise or information system will generate a lot of data or information. So the question is, how does one sort or even know what data is present within the dataset? The answer is quite simple, metadata! Metadata is very helpful, but what is it? Metadata is simply a data file about the dataset. It is literally, data about data! A metadata file is attached to a data file and can contain the following information: definition, data type, format, use, constraints, source, etc. There are many metadata files available for download, from software companies, or government agencies. Some of these templates will have many pages of forms to be completed, which are boring, and time-consuming to complete (which is why many interns find themselves with Q&A duties!). Metadata is essential, but creating or recording it is not a painful experience.

Metadata files can be customized to fit an organization's needs. Consider this scenario, and the organization decides that it needs only the necessary information about what information is in a dataset. All one needs is a name, item, and year of creation for the data. Metadata can double as the file's name "Streetsboro_StrPrjHist_2018." Streetsboro_StrPrjHist_2018 is the format "City_Data_Year created" where the City = Streetsboro, Data = Street Project History (StrPrjHist), and Year = 2018. A simplified form of metadata leveraging is having the file's name or title as a form of metadata.

Although it is very tempting to simplify metadata down to a single name, it is essential not to overlook the potential metadata has as a tool. One example is that one can provide a summary or description of the dataset into the metadata file. Metadata could be a quick one or two paragraph description of what is inside the data file. Consider the data as mentioned earlier Street Project History; a short description of the data might include that each street has a condition rating of either critical, poor, good, or excellent, thereby informing the user that this file will allow them to quickly determine which streets do not need repairs and which roads are in critical condition. A metadata synopsis can provide a quick evaluation to any user as to whether or not the data will be useful to their project.

5.5 System hierarchies

Every enterprise venture must have the system hierarchies established. Graham called these hierarchies enterprise structure data (Graham 2012). He explains that this data allows the system's activities to be reported, analyzed, by type of function or product grouping (Graham 2012).

TABLE 5.1

Hierarchy of WoodWeyr, LLC

WoodWeyr LLC				
Equipment		Hive Data	Supplier Data	Products	ID
Support	Woodenware	Weight	Blue Sky	Honey	1
Smoker	Hive Body	Feed	Walter T Kelley	Wax	2
Tools	Frames	Queen Located	Mann Lake	Pollen	3
Hammer	Inner Cover	Brood present	Dan Kaminiski	Propolis	4

Although this explanation is a little vague, we can clarify it a little further by considering the system hierarchy displayed with a table (see Table 5.1: WoodWeyr, LLC). WoodWeyr, LLC is the highest level of data hierarchy. Grouped under the WoodWeyr structure are the sub-hierarchies: equipment, hive data, suppliers, and products. Each of these sub-categories can contain detailed information. Consider the sub-category equipment and its sub-woodenware.

Woodenware in the world of beekeeping is the term for all of the components of a wooden beehive. Most of the beehive body parts (tops, inner cover, body, bottom) in the United States are pine or a type of hardwood. Roughly 50% of the frames that hold the wax and honey are also pine. Thus the category woodenware would allow someone to perform a search, analysis, to determine what type and how many individual hive components are on hand or deployed. The enterprise structure data provides the means to drill down further into the datasets for more detailed analysis.

Enterprise hierarchy data provides an organizational structure for the system's data and information. These hierarchy structures offer users a way to quickly determine what data is essential and how to find it. The structures also enable efficient querying of the data because each category (and sub) contains only one type of data for the company. Enterprise structure hierarchies provide organization to the system's datasets and groups.

There is another category of essential data found within the system, and that is reference data. One should be careful not to confuse reference data with metadata! Metadata explains or summarizes what is in the dataset (remember data about data!). Reference data is about how to organize the data found within a dataset or database. Typically, this data is seen as values in a look-up table or the world of GIS a domain value.

Domain values are an essential aspect of the geodatabase and are a line of defense against bad data from contaminating the geodatabase. Bad data can find its way into the geodatabase via user or editor error. A simple misspelling, transposing of letters, or a quick personal abbreviation (I will come back and add the proper notation later), etc. These are all data errors that can be easily prevented by the use of domains, otherwise known as reference data.

5.6 Data standards

Data standards establish the rules and procedures on how data is collected, defined, and recorded. Standards are necessary because they determine how everyone will share, transfer, and interpret the data. Standardized data can easily convert into easily understood information. Excellent, understandable, and easy to access information is the ultimate goal of the data architecture. The main objective of data architecture is to transform data into useful information.

Imagine your organization has purchased a new computer system. It now needs a memory upgrade, if every computer system is on the same requirements for memory then your update is relatively simple, and the IT staff goes out to the nearest computer store and purchases the proper size memory chip. However, if the organization has a non-standard memory chip, then it must search for the appropriate memory chip (the problem with proprietary hardware). A search for the proper hardware can take time and take the organization down the wrong path. If all of the computer requirements for memory were the same, it would not matter which company provided the memory chip, for it would still work within your organization's computers – the advantage of standardization either in hardware or data.

Now that we understand why standards are necessary, the question becomes, "How do we establish these standards?" First, we hold a meeting (yes it does sound like a cliché!) but you need to learn what people expect from the data! The standards will shape the data to meet what the organization expects from the data. If the data does not meet the expectations of the organization, then the resulting information will not help the organization. Standards are essential to people using or consuming data.

The standards for data must follow the same rules as the standards for the overall EA. 1. The data team must propose standards and they must be approved by the stakeholders. 2. Standards must be understood by everyone concerned with data collection, storage, or use. 3. All standards must be widely published and accessible by any data user or consumer within the organization. 4. Any established standard should have a review and expiration date; review and expiration dates will help to develop a practice of internal review that will help to keep standard fresh and up to date. 5. Remember to keep things flexible; don't feel that once approved a standard is set in stone forever; there is nothing wrong with changing over to a standard that is a better answer for the organization's needs (Bedford 2014a).

How do we determine what a data standard is? It is not only the architect who sets the standards for data, but it is also the field crews, users, and executives. Why do all of these people matter? Because they are the collectors and final users of the data! A data architecture connects the field version of the data with the final information consumed by the executives making business decisions.

Defining a standard can be confusing. One can use a pattern like those published by these organizations:

USGS National Geospatial Program Standards and Specifications
FGDC National Data Standards and Publications
FGDC National Standards Working Group
US Integrated Taxonomic System (ITIS)
US National Vegetation Classification (USNVC)

Data standards ensure that everyone is entering the same type of information (United States Geographic Service [USGS] 2016). Unfortunately, it is not a simple question on how to define a data standard. Definitions of a data standard run the gamut from amazingly complex to absurdly simple. The key issues that determine the success of a standard are: Was the establishment of the standard transparent? These issues will determine the success or failure of the data standards. Just like the overall EA, "buy-in" from the executives, stakeholders, and staff is critical for data standards to succeed. No matter how the data standard is implemented, without buy-in or support from the organization, it will be ignored, bypassed, or become irrelevant.

Data standards for any organization will only succeed if they have the full support of the organization's staff and stakeholders. This "buy-in" or complete support is crucial. The data collected in the field must be consistent with the recipient database. If the data standard is Bluebird Lane, then we should not find Blue Bird LN entered into the street name field. When the data collector does not support the data standard and uses Blue Bird Street (St) instead of LN, then an inconsistency will occur. This inconsistency might seem minor; however, when a query or report is performed based on LN, it is run against the database, anything with the abbreviation St is skipped. This omission will result in an incomplete report, which in turn adds an error to the final business intelligence analysis.

The last two sections defined data and demonstrated how data becomes valuable information. We also discussed the fact that data or information can occupy different levels within the organization. One of the most challenging issues in dealing with data is the disconnect that exists between people who create and those who ultimately use the data. A successful data architecture will enable everyone to understand the relationships between data creation, departments, and users.

One of the most straightforward tools for exploring these data relationships are reference cards (Joshi 2016). These reference cards provide a quick summary of a dataset. The information contained on these cards should be kept brief and easily understood. Table 5.2 is an empty template for data reference cards. Any data user gains a quick and accurate rating of the information's status. The card has the following information about the dataset: reference number, name, definition, owner, who uses it, alternate name, search and query

TABLE 5.2

Reference Card Template

REFERENCE NUMBER	X.X	I
NAME	File name of the dataset.	F
DEFINITION	What is the dataset or what it contains.	O
OWNER	Who owns or created the data.	R
ALSO USED BY	Which departments or users use the data.	M
OTHER NAME	Alternate name of the dataset (when imported).	A
TAGS(PREDEFINED)	What are the keywords for a query or search.	T
OBJECT SENSITIVITY	Special conditions that impact the use of the data (sensitivity, restricted use, etc.).	I O N
PROJECTION	Defines the map projection or datum (if there available).	S
FORMAT	Defines file format (.doc, .shp, .gdb, mdb, .xls).	T
LOCATION	Where is the data geographically located?	A
STATUS	Is the data complete, verified, in progress, archived, updated, etc.?	T U
INFORMATION QUALITY	Is the information good, trusted, old, not determined?	S

tags, sensitivity, map projection (if there is one), file format, location (city, state, GPS coordinates, etc.), status (in progress, complete, verified), and finally the information quality (good, bad, not determined, trusted, etc.). Note the information status bar is on the far right of the card. This bar can be color-coded to provide a visual reference that immediately shows data's quality (Joshi 2016). The color of this bar should reflect the information found in status and information quality: Green = Trusted, Orange = Restricted, Red = Not determined. This template is easily customized to meet the information requirements found in any organization's existing information system or archives.

The reference card shown in Table 5.3 is an example of a reference card for a street centerline for the City of Lyndhurst, Ohio. The reference number is 1.0, which means that this is the first card in the file. A greenfield on the right tells the user that the data is useful, updated, and verified, and can be trusted. All the information on the cards should be concise and follow the established data formats (formatting is a little later); therefore the name of the file follows the data architect's format CITY NAME + DATA NAME "LyndhurstStreetCenterlines." The definition is concise and right to the point. The department that created or "owns" this dataset is the GIS department, and the city service department also uses the data.

The dataset does have another name (Other Name), which is CityStreets. "Other name" might show that the dataset is an import from another source, or could have been existing within another system, before being included in the enterprise system. It is essential to keep the original name for the dataset because it maintains the path back to the original data if any questions or problems arise in LyndhurstStreetCenterlines.

Tags(predefined) are words that can be used to search or query the database to locate this particular dataset. Anyone who has performed a search

TABLE 5.3

Lyndhurst Street Centerline Reference Card

REFERENCE NUMBER	1.0
NAME	LyndhurstStreetCenterlines
DEFINITION	A line representing the very center of a street. These lines are divided into smaller segments from one intersection to the next.
OWNER	GIS department
ALSO USED BY	City service department
OTHER NAME	CityStreets
TAGS(PREDEFINED)	Centerline, Street, Lyndhurst
OBJECT SENSITIVITY	None
PROJECTION	NAD 1983, STATEPLANE OH, NORTH, FT
FORMAT	Feature Class
LOCATION	G:\\LyndhurstGeodatabase2016
STATUS	**Complete/Verified**
INFORMATION QUALITY	**TRUSTED**

on Google or Internet Explorer will recognize this as setting up a keyword search. Keywords or "tags" can help users locate data within a larger dataset. It is important to remember that users are reminded to input their keywords in the same format as the tag word. A simple misspelling of the word streets might pull up an entirely different dataset.

Object sensitivity is another way of discussing data security (Joshi 2016). This field contains notes about how sensitive the data is, what conditions exist before the dataset can be released to someone outside the organization. When the card refers to any data that is considered sensitive, this field should be highlighted and quick to draw the user's eye (Table 5.4). So the color bar should be assigned a unique color shade (remember green is good, and red means problem or bad). This author has always tried to stick with a shade of orange to represent sensitive data, it is not quite a yellow or a red, but it does catch the eye. No matter what color is chosen to mark confidential data, it should not be close to a green or red.

Projection is a field that is explicit to GIS maps, data, and systems. A map projection is a mathematical method of projecting the 3-D earth's surface on to a 2-D surface (a piece of paper). There are many different projections available to cartographers. However, if the projection of the data does not map the projection used by a base map, the data will be shown in the wrong map location. Providing the data's map projection will help to eliminate projection errors.

The field label format is where the data's native format is recorded. The most basic and universal GIS data format is a shapefile (.shp). Common file formats like MS Word, Excel, or Access extension can also be recorded here. It is essential that any UNUSUAL file format from open source, proprietary, etc. is entered into this field. This field is vital because it can determine whether or not the data's native format is compatible with the system's database.

The location field is simply the pathway within the system back to the dataset. The URL path is entered in this location. When using digital cards, then this path can be a hyperlink directly to the information. One must remember, if the information is marked sensitive or restricted, it would not be prudent to fill this field with a hyperlink. Status is the field showing if the information is complete, in progress, verified, etc.

The field for information quality is crucial because it quickly reveals what a user can expect from the data. When system users see the word "trusted" in this field, they can immediately assume that the data is accurate and up to date. In progress, incomplete, compiling, etc. are used to show that this file's data is not complete and it might not be ready for use. There are times when data enters the archives, but there is no way to check the information's accuracy – a situation where "not determined" should be placed into the field. Not determined informs users that the information accuracy has not been checked and might contain errors.

These reference cards have another purpose besides providing a summary of the data quality. They are building blocks for data mapping and architecture. The information in the fields can be used to create a data model (map). This data model will show the relationship between the data sources and users. Table 5.4 is a data model developed from the reference card 3.1.

One should expect a surprise or two when developing these card-based data models. It is relatively common to discover an unexpected user tapping into a data resource. The card, as shown in Table 5.4, reveals that most of the departments accessing the file LyndhurstPD_FDlocations are normal; police, fire, service, and planning. However, the health department is also accessing the information.

TABLE 5.4

Lyndhurst Flood Damage Reference Card

REFERENCE NUMBER	3.1
NAME	LyndhurstFloodDamage
DEFINITION	This is a file that shows the buildings impacted by recent flooding
OWNER	GIS Division
ALSO USED BY	GIS, Building, Planning, Health, Law, Fire, and Service Dept.
OTHER NAME	FloodDamageAssessment
TAGS(PREDEFINED)	Lyndhurst, Flood, FEMA
OBJECT SENSITIVITY	**File data cannot be shared with media unless authorized by Mayor, Fire Chief, or Police Chief**
PROJECTION	NAD 1983, STATEPLANE OH, NORTH, FT
FORMAT	Feature Class
LOCATION	G:\\LyndhurstGeodatabase2016
STATUS	**In Progress**
INFORMATION QUALITY	**NOT DETERMINED**

A health department accessing LyndhurstPD_FDlocations might be a surprise to system designers. However, this information is a starting point for a discussion with the health department about why they need the data, and how is it used? Perhaps the health department is tracking locations where EMT units responded to victims who suffered heart attacks, strokes, or accidental injuries (broken bones). Public health officials might be trying to determine if public safety can improve through a reduction in dangerous street crossings or hazardous locations. A meeting between public health officials, police, etc. might be required to determine if this information is necessary for public safety and if there is a better way to collect or disseminate it to the department.

These cards can be used to determine if anyone had previously established data standards, principles, or formats. The field NAME would be a perfect indicator if any data principles or formatting conventions were applied to the existing data. When the file names consistently follow a format like "name of city_datatype_year," then someone probably understood the benefits of establishing organization within the datasets. Unfortunately, most of your legacy datasets will have names that bounce from John's_points, Streets, input_output, etc., which indicates that the data is not organized. These reference cards can provide a quick estimate of the data quality and level of organization.

Enterprise GISs often warehouse and import data from many different sources: multiple departments, online, federal, state, local governments, etc. Remember, an enterprise GIS is used by staff within various departments, skillsets, professions, and each one will have a unique need for data analysis. A common issue within enterprise information systems is that data/information users often do not have any idea of how to create data and the creators of the data have even less of an idea who is using the data/information (Graham 2012, p. 91). Graham considers this issue to be the "crux of the problem" that must be overcome by practicing Master Data Management (MDM).

The heart of a GIS is data, and its beat is the number of projects that successfully convert data into valuable information. The primary function of GIS is to place data and map it over the earth's surface. It is critical that the data heart of a GIS be clean from errors or inaccuracies. Preventing mistakes is the first line of defense when protecting GIS data from errors.

When waging war on data errors, everyone should use all of the tools at their disposal. GIS system designers, developers, programmers, etc. should rummage through the toolbox of the data architect. Data architects are required to ensure that datasets are clean, secure, and readily available. Their toolbox has ideas and methodologies that can have a positive impact on the data in an enterprise GIS.

The first tool that should be "borrowed" from the data architect is the concept of developing data standards. GIS users are very proud of the fact that there are many ways one can arrive at an answer to a question. Unfortunately, this means that there is no standardization of how to format, store, or use data within the system. Each person will follow what they think is best, creating a dataset that is filled with errors, duplication, and is disorganized.

Establishing a set of standards for the GIS dataset is not very complicated and will help to provide an efficient and streamlined flow of data into and out of the enterprise system. Data standards will establish definitions for each piece of data and how it should be formatted. Standardizing data procedures will prevent the loss or misplacement of data and the resulting project delays or confusion that occur when information is missing.

A foundation stone for both enterprise architecture and GIS is the creating of a central warehouse of data and information. This concept boils down to creating a folder or database that is accessible for everyone on the system. A geodatabase can act as a data warehouse for the enterprise GIS. Data warehouses create a central location for all system information. These geodatabases provide database capabilities for data distribution, versioning, status, lifecycle, etc.

A geodatabase has two main structures: feature datasets and classes. A feature dataset is similar to a windows file folder because it is used to house a group of feature classes that share a common theme. Feature classes are collections of similar items which are a point, line, polygon or text. Consider this example. A geodatabase for the City of Strongsville has the following feature datasets: LAND, TRANSPORTATION, GENERAL. Data about the city's streets, railways, and pavement conditions is a separate feature class that is grouped and stored in the feature dataset TRANSPORTATION. A feature dataset size is limited to 1TB, which would seem like a limitation until one realizes that a geodatabase can contain an unlimited number of feature datasets. Theoretically, an unlimited number of feature classes means that a geodatabase can grow to almost any size without sacrificing performance!

Since the geodatabase can contain an unlimited amount of data, it has the potential to carry the same amount of data errors! Data standards can be used to prevent or reduce these data problems. Naming standards for folders and files when applied to the feature datasets and classes allow the user to identify where and what something is quickly, and the date the information data was added. Now a simple glance at the database will let the viewer see what is available and the date of data entry.

Feature classes show data by one shape (point, line, or polygon). All the information for a form is in an attribute table within the feature. These attribute tables consist of two or more fields (OBJECTID, HOUSE NUMBER, STREET, etc.) for the data about the items in the class. Establishing and enforcing data format standards will prevent data entry errors as data enters the table. Format standards are the first line of defense for the geodatabase.

5.7 Reference data

This section is where the concept of reference data meets reality. Databases commonly leverage a "look-up" table to help provide information or fill information in another table, particularly for a data entry application on a mobile device. These tables are pre-loaded with data values that can be used to validate entry values; if the data entered is part of the list, then the entry is accepted. If the entered data does not match a value in the table, then the entry is rejected; and the user is requested to make another choice. Look-up tables are a useful technique for preventing the wrong data entering a cell.

The geodatabase can create and store these look-up tables and make them available to any feature class table. These are called geodatabase domains and are very useful to define and enforce data format standards. Domains can be set to one of several data formats (string, double, float, large integer, etc.) and consist of a code and description column. The code column is where the code or abbreviation for the data and the description defines the code value. An example would be the domain PIPE TYPE. PIPE TYPE is the value of the text, and the CODE value contains an abbreviation for pipe material, and the DESCRIPTION is the full word.

Geodatabase domains are easily updated or created with MS Excel tables and the "table to domain" tool. This tool creates a domain from any MS Excel table with a two-column format, defined as CODE and DESCRIPTION no matter how many values it contains. The data undergoes a quality check for typos, spelling, or proper format (i.e., date = M/D/YEAR) before being uploaded into the domain. Domains are a tool that should be used to prevent and protect the geodatabase from data entry or format errors.

5.8 Data formatting

There are many different formats for data. It is effortless for people to "personalize" data formatting by simplifying using whatever format is comfortable. Personalized formatting is okay if one is only working on their data system. However, when one is working in a system that is accepting data from multiple users or sources, individual choices about formatting can become a significant issue.

There are many different formats for data. Consider that there are several choices for formatting a simple date: 3/14/2016, 03/14/2016, 14-Mar, Mar-2016, 2016-Mar-14; and those are just a small selection of the available formats. Names and abbreviations are other examples where there can be a thousand different choices! Everyone has a favorite abbreviation or acronym for a word, title, or place. Personal preference becomes a problem when multiple people are entering data into the same database.

Datasets often include many basic errors such as typos, incorrect formats, entry errors, etc. Architecture standards will help to eliminate these errors by creating a formalized set of rules that force the standardization of data creation and use (United States Geographic Service [USGS] 2016).

Establishing a set of data standards is essential to getting everyone on the same "data page" when communicating ideas or developing applications. The lowly date attribute is an excellent example of how a data standard can be applied.

5.9 Geodatabase domains

The question becomes, how do we control what enters into an enterprise GIS? The answer is to work with the geodatabase as domains. The domain feature of a geodatabase seems to be forgotten by users, programmers, and designers of GISs. However, this is a mistake because domains can be handy tools for controlling and eliminating data input errors. The secret to successfully leveraging domains as a data protection tool is the setup.

A geodatabase domain is in the most straightforward explanation a dropdown list of input choices. There are two ways to create a domain in a geodatabase; first, a person can create a new domain list and enter each of the options found on the list individually. Manual entry is not a wrong choice if the list of data choices is relatively small (5 or 10 choices); however, it is not a very efficient choice if the list of options is much more extensive, say over hundred input options. Keeping an updated list of one hundred or more data input choices can be tedious and labor intensive. Not only that, the domain list cannot be made to reset in alphabetical order. The newest domain item appears at the bottom of the list.

There is another way to create geodatabase domains, and that is through Microsoft Excel (Table 5.5). Creating a new domain is not a complicated process, but it does require attention to detail. Again, it comes back to the setup; the proper set up of the table is key to a successful import of the table into the geodatabase domain. There are only two fields that one has to set up for a proper import from Excel to the geodatabase.

The geodatabase domain has two fields: code and description. Code is the column where value represents the data input. The description is simply the column that defines the value placed in the code field. Consider this example: Code = CONC and the corresponding cell in Description would

TABLE 5.5

Geodatabase Domain for Pipe Material

CODE	DESCRIPTION
PVC	Polyvinyl chloride
CONC	Concrete
RCONC	Reinforced Concrete
CI	Cast Iron

be Description = CONCRETE. The table entry would look like this CONC | CONCRETE. Domains seem complicated, but in reality, coded domains are relatively simple.

Your Excel table should mimic the domain! Two columns, one labeled CODED, and the other DESCRIPTION are filled out the same way as above; however, there are two advantages to setting up a domain in Excel and then importing or updating the geodatabase. First, in Excel, one can sort the columns, as opposed to the static domain where the last item entered is at the bottom. Second, it is easier to add or replace one or two lines in the Excel table and then upload the changes to the geodatabase. Remember, the goal of using domains is not only to protect inputted data from errors, but it also makes things more comfortable for the person inputting the data.

5.10 Summary

Data architecture is a sub-architecture of the overall EA plan. However, it can be a stand-alone strategic plan that deals with how data is stored, used, and converted into information. It is an essential piece of the EA because the data architecture is the primary line of defense against bad data corrupting the system's information set (database).

5.11 Assignment

Assignment 5: Data examples

This chapter has defined the difference between data and information. So can you tell which of the following is a piece of data or information? Water, Scotland, 6, 58 Degrees F, Main: Please explain why in a few short sentences.

Bibliography

Bedford, D. 2014a. *Business Architecture*. Kent, OH: Bedford, Denise Ph. D.
Bedford, D. 2014b. "Business On A Page (BOAP)." *Business On A Page (BOAP)*. Kent, OH: Kent State University, Dr. Denise Bedford.
Dueker, Kenneth J., J. Allison Butler. 1997. *GIS-T Enterprise Data Model with Suggested Implementation Choices*. Portland, OR: Center for Urban Studies, School of Urban and Public Affairs, Portland State University, October 1.

ESRI. 2007a. *Enterprise GIS for Local Government*. Redlands, CA, December. http://www.esri.com

ESRI. 2007b. *GIS Best Practices Enterprise GIS*. Edited by ESRI. Redlands, CA, January 1. http://www.esri.com

ESRI. 2015. *Enterprise GIS. GIS Dictionary*. St. Louis, MO: ESRI, Environmental Systems Research Institute.

Gartner, Inc. 2008. "Gartner Clarifies the Definition of the Term 'Enterprise Architecture'." *Gartner Research*, 15. August 12. doi:G00156559

Gill, A. 2013. *Defining a Facility Architecture within the Agile Enterprise Architecture Context*. October 1. doi:WP0107

Graham, A. 2012. *The Enterprise Data Model: A Framework for Enterprise Data Architecture*. San Bernardino, CA: Koios Associates Ltd.

Joshi, P. 2016. Lecture Data Architecture. *Data Reference Cards*. Kent, OH: Kent State University, Parasanna Joshi, February 18.

Labuschagne, L. 2011. *Building Enterprise Architectures for Non-Architects*. September 1. doi:WP0011

Lapkin, Anne, Philip Allega, Brian Burke, Betsy Burton, R. Scott Bittler, Robert A. Handler, Greta James. 2008. "Gartner Clarifies the Definition of the Term 'Enterprise Architect'." *Gartner*, August 12: 15.

Michigan Department of Information Technology. 2007. *From Vision to Action: Enterprise Architecture – Strategic Approach*. Michigan Department of Information Technology.

Ross, J. P. 2006. *Enterprise Architecture as Strategy: Creating a Foundation for Business Execution*. Boston, MA: Harvard Business School Press.

Rouse, M. 2007. *Enterprise Architecture (EA) Definition* (TechTarget), June 07. Retrieved October 2015, from SearchCIO: http://searchcio.techtarget.com

Schulmeisters, K. 2014. *Enterprise Architecture: Bridging Entrepreneurs and Hard Problems*. November 1. doi:WP0168

Sieber, R. 2000. "GIS Implementation in the Grassroots." *URSIA Journal* (Vol. 12, No. 1), 15–29, Winter.

The Open Group. 1995–2015. *Enterprise*. The Open Group. January 1. www.opengroup.org/subjectareas/enterprise

Thomas, C. 2009. "What's Your Definition? Looking at What Enterprise GIS Really Means." *ArcUser*, 3032. http://www.esri.com

United States Geographic Service (USGS). 2016. *Data Standards*. http://ww.usgs.gov/dtamanagement/plan/datastandards.php

Wilson, North Carolina. 2012. *A GIS Manager's Dialogue: Is Your Business Semi-Integrated or Fully System Integrated?* NC.

6
The EA/Enterprise GIS Toolbox

6.1 EA/EGIS tools

This chapter introduces a set of standard tools that are available to anyone designing an enterprise architecture/enterprise geographic information system (EA/enterprise GIS). The "tools" covered are Business on a Page (BOAP) that encourages brainstorming sessions for sketching out an early vision of the system. A business capability model that will help to determine what the business produces. The National Association of State Chief Information Officers (NASCIO) Maturity Model will help designers determine the success and growth of the system. Finally, Heat Maps, Workflow models, UML activity diagrams, along with the strategic analysis methods of Strengths, Weaknesses, Opportunities, and Threats (SWOT). All of these tools can be "borrowed" from the enterprise architect's toolbox and used by GIS designers.

This chapter will offer an introduction to the EA tools used by the author for creating enterprise GISs. These methods can be used by enterprise GIS designers without requiring any modification for transforming the item into a "proper" enterprise GIS tool. Overall data is data, and the tools or methods used by one field (enterprise or data architecture) is vastly different than GIS data; therefore a person cannot possibly use the ideas from EA to manage GIS information. The author would like to point out that the only difference between geo-spatial dataset and "regular" dataset is that the geo-spatial dataset will contain a geoidentifier that can place the data on the earth's surface. The geoidentifier can be an address or a set of GPS coordinates, both of which are numerical data. This means that data is data, and people can use any of the tools available to manage enterprise GIS data.

The enterprise architect will guide the organization through the process of designing, implementing, and maintaining an organization's enterprise architecture. Designing enterprise architecture is a process of listening, discussion, and compromise. It is essential that every level from the level of executive management down to the newest member "buy-in" assumes ownership of the architecture (Sieber 2000). Successful architecture is used and supported by all levels of the organization. The EA/enterprise GIS toolbox has many tools that can help a system designer track the progress of the system's growth.

6.1.1 Business on a page (BOAP)

A BOAP can be a starting point for business architecture; it is a high-level look at a business or organization. It is used to create a representation of the company that will make sense to the business's stakeholders (Bedford 2014b). The BOAP is also an anchor point for aligning the business's resources. This plan is usually created by brainstorming; it is not meant to go into in-depth detail. It will undergo several incarnations before reaching a "finished" state, and it can always be "tweaked" at a future date. A BOAP is used for determining what architecture segments are within the organization and creates a business framework for creating a consensus on what areas the enterprise architecture team should concentrate their attention on and assess the alignment between application needs and planned initiatives (Bedford 2014b). The information in a BOAP is dynamic and should be expected to change not only during the initial design phase; this document changes as the company matures.

The development of a BOAP should not be a stressful exercise. BOAP is a practice in brainstorming or quickly developing ideas. Everyone should feel free to throw out their concept and see "if it sticks!" Brainstorming concepts is where the development team can run wild with ideas. No idea should be discounted or dismissed in this process. The BOAP is all about searching out and considering new ideas, regardless of how outlandish it sounds. One never knows what concept might be the one to get the ball rolling on how the business should develop.

The BOAP (Figure 6.1) is divided into three areas: strategic, operational, and enabling. The strategic portion has typically two fields that define the company's business (Bedford 2014). The illustration shows that the new SBE Cyberdragons will focus on the areas of GIS services and GIS consulting. These areas of activity are the main focus of the company. The rest of the BOAP deals with the understanding that everything should support or fall into one of these two categories.

Just under the strategic level is the operational level of the BOAP. The operational level has two distinct groups. The first group is the set of "balloons" just below the strategic level. Please note that these operational groups are under one of the two strategic concerns. Consider Figure 6.1. The items' internal products and programming are squarely under GIS services. The dots extend the primary group GIS services over to the middle category geographic information systems.

Geographic information systems are the middle category since it has the dual roles of providing a service or being a product of consulting. The intermediate category is where any "hybrid" or item is characteristic of the two strategic objectives. It is quite possible that more than one of these sub-categories would cross over between the strategic groupings, but the hybrid

The EA/Enterprise GIS Toolbox

CHAGRIN VALLEY ENGINEERING GIS (BUSINESS ON A PAGE)

GIS SERVICES ──────────────── **GIS CONSULTING**

Strategic

Operational

INTERNAL PROJECTS	PROGRAMMING	GEOGRAPHIC INFORMATION SYSTEM	ADVISING	TEACHING
- DEVELOP PROJECT PARTNERSHIPS WITH CVE DEPARTMENTS	- WORK IN PYTHON, VBA, & DOT NET	- FIND OPPORTUNITIES FOR CONSULTING.	- PURSUE OPPORTUNITIES FOR CONSULTING.	- PROVIDE THE BEST INSTRUCTION FOR GIS
- DEVELOP A NEED FOR GIS INPUT, DATA, ON COMPANY PROJECTS	- WORK WITH OPEN SOURCE	- CREATE RELATIONSHIPS WITH MUNICIPALITIES	- PROVIDE THE BEST ADVICE & OPPORTUNITIES FOR CLIENTS	- PROVIDE THE BEST INSTRUCTION FOR DATABASE/WEB PROGRAMS
- DEVELOP A GIS VIEWER SYSTEM FOR CLIENT CITIES & OFFICE STAFF	- DATABASE & GIS MODULES	- DEVELOP DATA/INFORMATION SHARING AGREEMENTS WITH COUNTIES	- ENCOURAGE COOPERATION OTHER COMPANIES	- PROVIDE THE BEST MENTORING FOR INTERN/CLIENTS
- INITIATE APPLICATION DEVELOPMENT	- DEVELOP DATA MODELS	- RESEARCH & PURSUE GRANT FUNDED OPPORTUNITIES	- PARTNER WITH OUTSIDE COMPANIES ON ADVISORY BOARDS	
	- DEVELOP MOBILE APPLICATIONS FOR FIELD WORK			

Enabling

1. ENCOURAGE PROJECT COLLABORATION
2. ATTRACT AND RETAIN TALENTED STAFF
3. ENCOURAGE EXTERNAL PARTNERSHIPS
4. ENCOURAGE ASSOCIATION MEMBERSHIP
5. SUPPORT CONFERENCE ATTENDANCE & PARTICIPATION
6. FULFILL ALL CONTRACT AND LEGAL OBLIGATIONS
7. MANAGE COMPANY DATA AND MAPPING RISK
8. MAINTAIN HIGH SKILL LEVELS
9. PROVIDE A NURTURING WORK ENVIRONMENT
10. PROVIDE A GOOD BENEFIT PACKAGE
11. MANAGE FINANCIALS
12. MANAGE INFORMATION
13. BE A GOOD COMMUNITY MEMBER

FIGURE 6.1
Example of a BOAP.

categories are the natural break or transition between the two strategic groups. We will take a closer look at a hybrid category in just a minute.

These categories or ideas are sub-groupings and are broad ideas of what the company GIS division could provide internally as a GIS service or externally as a consulting service. Referring back to the illustration, note that each of these sub-categories sits above a bubble in the operational field. The operational "bubbles" are the items that support the sub-categories. The ideas in the operational bubbles are items or ideas that support the above categories.

These operational areas are ideas that support the category above them. Consider the category "programming," the first statement is "work in python, vb and dot net." This statement is defining the areas or languages that any "programming" development for applications or applications occurs in one of those languages. The next statement is that the company will also work with any open source program or programming language. These two statements, if followed or supported by the employees, will enable the company to fulfill the requirements of programming. The idea with the operational ideas is each statement in the operational category will support the strategic sub-category above.

The enabling section of the BOAP should have 13 or 14 short statements. Each statement is an idea that will enable people to accomplish operational and strategic initiatives. Consider statement number six (above example BOAP): "Fulfill all contract and legal obligations." This statement is quite merely reminding everyone to fulfill and follow through on promises. When a company or its employees don't live up to promises or obligations, not only do customers lose faith, but internal employees begin to question the value of the company's ethics. This type of situation will lead to a gradual loss of reputation and valuable employees, crippling the efforts to meet the strategic objectives for GIS services and consulting.

The category of geographic information systems is in the center of the second strategic and operational sections. GIS is an example of the aforementioned "hybrid" category. Designing and implementing a GIS would fall under the category of GIS services. There is much more to developing geographic information systems than only unboxing computers and installing the software.

The line between GIS services and consulting becomes blurred during a relationship developed between the client and company staff. Advising and educating clients about GISs and their capabilities moves the category into the realm of consulting. Any new system requires that client staff need training on how to use their new tool. Training materials, classroom, and field instructions, all fall under the item of education, another branch of the consulting initiative. Developing a custom geographic information system will require the concepts of GIS services and consulting.

It is essential to realize that a BOAP is an exercise in brainstorming that will require many restarts, changes, and sometimes "passionate" discussions before a finished product emerges. BOAPs are tools that discover the business's (or department's) goals, concepts, and how to meet the objectives (Bedford 2014b). Each BOAP should be quickly filled out, with no regard for sensibility or reasonability. Throw the idea out and see how it might fit into the rest of the BOAP. Brainstorming is about developing new ideas, not immediately throwing ideas away.

Do not be surprised or anxious if a BOAP for a business, department, or new organization goes through many discussions and planning sessions. A common theme will begin to emerge as people discuss and share ideas. These commonalities are the seeds for the "final" BOAP. Eventually, these common ideas outweigh the differences. When the majority agree on the BOAP's content, then you can consider it ready for presentation. Even finalized BOAPs will change, never be afraid to go back and alter the BOAP if a new idea is better suited to the plan.

6.1.2 Workflow model

Workflow is a model that is created to describe the tasks, decisions, inputs, outputs, people, and tools that are involved in a specific company process (Schedlbaure 2016). These models demonstrate HOW! (HOW each process task interacts with another job, HOW information flows through the process, and HOW workers are involved in the activity). A workflow model can be beneficial in determining potential areas for bottlenecks, duplicate effort, or redundancies that might lead to or cause work interruptions, delays, or stoppages.

Workflow modeling is a very challenging activity; however, it does provide a great value when done correctly. These models can make a complicated process easily understandable to anyone. They can effectively communicate people's roles and responsibilities within system processes. Communication prevents overlap of responsibility, effort, and the duplication of processes. Workflow modeling will allow stakeholders to quickly and easily understand complex system processes, which enables them to have an appreciation of the system's value.

6.1.3 UML activity diagram

UML was developed by the Object Management Group (OMG) as a method to standardize complex modeling systems. UML is an acronym that stands for "Unified Modeling Language" and is the language used to standardize how architects visualize the architectural "blueprints" when creating a system architecture. Business activities, processes, database schemas, are displayed in models that are in a single and widely used modeling language. Using the UML makes it much easier for the architect to share the architectural model with people inside and outside of the business (Schedlbaure 2016).

The author has found that these activity diagrams are very useful in determining who else will be accessing a data point or impacting the system processes. UML activity charts do a nice job of displaying how information travels through a system. The UML charting should be drawn after the basic system design has been laid out. Otherwise the system designer will tend to overthink or overcomplicate the system with unnecessary lines and arrows.

6.1.4 Strategic analysis methods

These methods are Strengths, Weaknesses, Opportunities, and Threats analysis and a value chain analysis. The SWOT analysis determines what the main strengths and weaknesses of businesses are; it also predicts what threats could hurt the company and where opportunities exist in the markets.

A SWOT is similar to the BOAP in the fact that they are both exercises in brainstorming. However, the SWOT looks at specific items about the business. The best way to view a SWOT is to create a four-quadrant table. Place an S in the upper left, W in the top right, O in the lower left, and T in the lower right. Have the stakeholders or decision-makers (without a discussion) quickly write down on separate Post-it notes all of the organization's strengths, weaknesses, opportunities, and threats. (One option is to have them include a description of no longer than three lines.) When they finish, post all of the Post-it notes into the proper category.

Now have everyone review the Post-it notes by category. List ideas as starting from the highest priority to the lowest. When there is a case of the notes with the same idea, that particular idea should be moved up in the rankings, when more than one person has noticed that idea or concern. The purpose of this exercise is to help develop a business strategy to take advantage of the company's strengths, improve weak areas, exploit opportunities, and above all protect itself from threats. A SWOT analysis can be and is very useful in developing an enterprise architecture.

The value chain analysis helps to determine what specific activities can give the company a competitive advantage and how these can build the company's value in business markets. This analysis is made up of three components: primary activities, support activities, and margin.

Primary activities are basics of the business take-in items, services, etc. and the output is to provide these items, services, etc. to customers and clients. Support activities are items like technology, human resources, and infrastructure that promote or aid the primary activities. Margin refers to the company's profits from successfully executing these three activities. Successful companies reach a balance where the primary and support activities work well together to lower costs, produce more revenue, thereby resulting in higher profits.

Let us take a look at a value chain analysis for a GIS firm. The primary activities are owners and executives who are concerned with business operations. Project developers find new business opportunities and maintain client relationships. GIS managers and coordinators ensure client satisfaction.

The support activities are GIS personnel who are responsible for providing services to clients – technology staff who are trusted to make proper investments in technology to support the primary activities. The administration staff is responsible for daily operations and interfacing with deliveries, office management, and miscellaneous items.

The margin (or profit value) is at 6% after expenses. Management is trying to increase this profit margin by focusing the business operations on specializing in providing interactive web maps. This strategy is called specialization, where the company becomes very specialized in one unique skill and can charge a premium to clients. However, to ensure that this strategy is sound, the project developers will perform market research to evaluate opportunities and competition.

6.1.5 Backcasting

Backcasting is a method of forecasting (working) backward from an envisioned future event. One declares a next objective or goal (like landing on the moon) and then works backward from this point to determine what processes, materials, steps, etc. will be needed for the team to accomplish the future goal (Schroede and Tilley 2016). This method allows people to develop strategies and procedures for achieving a goal by working backward from the next point or purpose.

A classic (and probably the best known) example of backcasting is the Apollo program that successfully landed men on the moon. President Kennedy tasked NASA with successful landing a man on the moon and then bringing him safely home. NASA engineers looked at the mission from the moon back to earth. Each step to get back would be mirrored with a step to go from the earth. History recorded Neil Armstrong's successful stepping out onto the moon, which was the result of a daring backcasting project.

Let us look at a more down-to-earth example of backcasting. Consider that you are on a committee planning a two-day GIS convention. The first item on the agenda is when to hold the conference. Dates for this convention will depend upon facility availability, calendar proximity to other events, and the desired time of the year. However, the committee will need time to plan, recruit presenters, and vendors for the convention (also known as the show).

Providing the appropriate amount of time for planning a convention is the first step in the planning process. Convention planning requires at least one year (12 months) of detailed planning. Therefore, if you start planning in January 2019, the convention should occur at the earliest in January 2020. However, January through March are potentially the worst months for winter storms in many areas of the United States. The decision is made to move the convention into the Spring months.

May is the chosen month, and now the planning can begin in earnest. First, a facility (hotel, convention hall, etc.) is checked to see what dates are available.

Let us assume that the dates May 9–10, 2020, are available, and the committee reserves these two days. Now it is time to work back toward the current month of January 2019.

The big item for the show is content. Without content or presentations, no one will want to attend or sponsor the convention. Not only must "a call for papers" must be made, but time must be allotted to review and chose the presentations for the conference. Remember, the convention's content will impact who attends, which vendors will sponsor or exhibit at the show.

The call for presentations or papers will go out shortly after signing the contract for space with the host facility. Thus, if the show is May 9–10, 2020, the notice requesting submission presentations opens June 1 and closes on July 31, 2019. Now, from August through September, submissions can be reviewed and selected for presentation at the conference. These presentations that made the "final cut" make up the convention's information content.

Once the convention's presentations are selected, and the corresponding abstracts listed on the website and marketing materials selling space in the vendors' area and show sponsorship can begin. The success of recruiting vendors and sponsorships for the convention is finalizing the show's dates and content early. According to the timeline, content and the convention dates can now be posted online in late September or early October.

A goal for the person recruiting exhibitors (vendors) is to sell exhibit hall space before the opening of the show. Remember, the show dates are May 9–10, the exhibitor recruiter should cease selling and switch to planning the exhibitor hall's final layout six weeks before the show. Pre-planning the exhibit hall layout ensures that a hall map will make it into the show's onsite program and listed on the website as a marketing aid for attendee registrations. Therefore, mid-March is the cutoff for recruiting vendors, exhibitors, and sponsors.

Planning and executing a convention is a challenging undertaking and will require many hours of planning and effort. Each convention area registration, vendors, sponsors, presenters, presentations, etc. are interdependent. Consider this, if the majority of abstracts seem mediocre, potential exhibitors, sponsors, and attendees will conclude that this show's quality will be average or even below average. The perception that the show is mediocre results in low numbers of attendees, sponsors, and exhibitors. Low attendance means lower revenue and possibly the convention losing money. The technique of backcasting allows planners to determine the when and how departments interact, potential issues, and what is a reasonable timeframe for completing the project.

Case of the City of Strongsville

We have discussed backcasting as a strategy with an example of setting up some type of convention or conference. Let us consider backcasting as a strategy for a municipality's enterprise GIS. The previously mentioned City of Strongsville enterprise GIS project is an excellent example of the backcasting technique. Chagrin Valley Engineering (CVE) was selected as the project consultant. The author was CVE's primary GIS coordinator and system designer for this project.

The city specified that the project had to start in April and finish by mid-December 2013. CVE was contracted to finish and turn over the system to city employees on December 20. This also included a three-week introduction to GIS and how to work with their enterprise GIS. CVE entrusted the project's planning, build, implementation, and training to the author.

The first stage of the planning was to determine a timeline for the project build with the backcasting technique. First, the date December 20, 2013, was noted as the "day of turnover" and the city assumes total control of the enterprise GIS. The project required a "buffer" of days as protection from problems. A protection buffer is important to a city-wide project because it provides the contractor with time if a serious problem should arise. These "extra" days will give the team a little breathing space to fix a problem, while maintaining production and keeping the work on schedule. The author settled on a five-day buffer, so he noted that December 16 would be the day of turnover.

The contract requested a training period for training city staff on enterprise GIS concepts, methods, and ArcGIS desktop. Fortunately, a training period could run right up to December 20, and the "graduation gift" for the class would be assuming control of the city's enterprise GIS. Training materials could be developed and the classes conducted during the final weeks of the system build. The author considered December 20 was the last day of the training and December 2 was selected as the beginning of a five-session training class. These sessions were spread out over the training period to allow city staff to handle normal job responsibilities, practice the lessons, review materials, and ask questions.

Time had to be set aside for developing the classes and training materials. CVE's contract required that the training cover introduction to GIS, mapping, digitizing or querying data, and ArcGIS desktop. The city had requested a "high-level" class with emphasis being placed on digitizing, querying, and collecting data. Even with a high-level view of GIS, mapping, and data management, developing the training materials was estimated to take four weeks. The author decided to devote the entire month of November to developing and finishing all of the materials required for a training course.

This project also had five main tasks that had to be accounted for in the final timeline. Task 5 was the delivery and installation of the final product. Task 4 was to develop a digital utility map and training in GIS. The main part of the project all geo-referencing and digitizing of data was found in Task 3. Strongsville was gifted with a wonderful asset of a complete pdf library of all construction documents for all work in the city. Organizing and inventorying this document archive was Task 2. Task 1 was a pilot project to ensure that CVE's methodology would meet the city's accuracy specifications.

The author hopes that the reader noticed that all of these tasks have been listed in descending order. Listing items in descending order helps the designer work backward from the final task that ends the project, to the first task that starts the project. Task 5 was allotted three days for the delivery and installation of the final system. Three days included a one day buffer for problems, one day to coordinate with IT, and one day for system installation, for a grand total of three days.

Remember the project build was expected to end on December 16 (Monday), so the installation should begin on December 11.

Task 4 was constructing the overall utility map of the city's storm, sanitary, gas, and water information. There would be a digital and hardcopy (paper) version of this infrastructure map distributed throughout the city. Since the digital map would actually be constructed through the digitizing performed in Task 3, Task 4 was really a quality control check. Task 4 was allotted one week of time for completion. This means that Task 4 should start no later than December 2 and finish on December 6, 2013.

The "meat-and-potatoes" of this project was Task 3. Geo-referencing and digitizing the city's infrastructure into the overall geodatabase was the most important and longest running part of the system build. Task 3 was expected to take three and a half months to geo-reference and digitize the city's infrastructure. CVE's team digitized a staggering amount of infrastructure: 205 miles of sanitary sewers, 227 miles of storm sewers, and 183 miles of water mains. The team also digitized 2,862 fire hydrants, 9,322 manholes, 5,014 catch basins, 1,567 yard drains, 848 headwalls, and 657 stormwater outfalls, during three and a half months. Task 3 was set to end on November 1 and start on June 10, 2013.

Task 2 creating the document inventory was also a monumental effort. All of the digital documents and construction plans were placed in an archive. The digital archives contained over 5,000 individual sheets for construction projects. Team members individually reviewed each sheet and determined what information was on the sheet that should be digitally recorded into the database. This process of sorting, inventorying, and identifying infrastructure information was allotted four months. Task 2 would end on June 30 (a slight overlap with Task 3) and start April 15, 2013.

The last item to be considered, but the first part of the project is Task 1 the pilot project. This pilot project was restricted to two side streets. A pilot project proves or demonstrates the consultant's methodology is sound, establishes working relationships with staff, and provides immediate feedback that something might be an issue that impacts the project deadlines. CVE was provided with two weeks to complete the pilot. This is entered on the project timeline as ending on April 12 and starting on April 1, 2013.

A project's timeline can be displayed as a graph, a chart, or a table. A simple listing like the one below can also be used:

City of Strongsville enterprise GIS build timeline

City's Project Start date April 1, 2013 End Date: December 20, 2013

Consultant's Project Start Date: April 1, 2013 End Date: December 16, 2013

System Training Start Date: December 2, 2013 End Date: December 20, 2013

Training Materials – Start Date: November 1, 2013 End Date: November 29, 2013

Task 5: Delivery and Install – Start Date: December 11, 2013 End Date: December 16, 2013

Task 4: Utility Map Development – Start Date: December 2, 2013 End Date: December 6, 2013

Task 3: Geo-Referencing and Digitizing – Start Date: June 10, 2013 End Date: November 1, 2013

Task 2: Document Inventory – Start Date: April 15, 2013 End Date: June 30, 2013

Task 1: Pilot Project – Start Date: April 1, 2013 End Date: April 12, 2013

The author believes backcasting is a useful technique for planning an enterprise GIS. The enterprise build team determines system size, departments supported, data storage requirements, and when it will be operational. The planners work backward from the expected operational date to establish a timeline for a system build. Backcasting will reveal how and what, when goals or processes, should be started or achieved.

6.2 Present/future state model

The present state or current state model will provide a snapshot of what the system's status is today. It attempts to realistically portray how the system is operating in the present day; there is no attempt to show future expansion or applications. There should not be any attempt to hide or gloss over missing, underperforming, or incomplete sections. A present state model must be truthful and accurate.

A future state model uses the present state model as the foundation. The future state model reveals what steps are necessary to achieve the present state model (Graham 2012). When one creates a model of the future state, he or she is creating a strategic vision about what the system will become. A future state vision can be used as a guide or checklist to determine if data, applications, procedures, etc. should be added or deleted from the system.

The criteria are simple: "Does this help or hinder the system from reaching a future state goal?" (Bedford 2014). Once the system has achieved its planned future state, then the future state becomes the present state, and a new future state envisioned. This recurring cycle of growth ensures that the system keeps pace with the organization's growth and needs.

6.3 Process model

Processes are all about what the organization does or what occurs during business operations. This model is an in-depth examination of the procedure completion, not who is responsible for finishing the procedure. Each organization process is unique and is defined by a set of actions that relate only to it (Bedford 2014). Completing the processes will help the organization achieve its business goals. Process models display what process is required or finished before a task is considered finished.

Process statements follow the simple phrase structure of verb and noun (Build Geodatabase), and there are two levels of process statements (Baer 2014). Level 1 process provides an overview of categories that are general headings. A level 1 process could be Build Geodatabase, and this would be concerned with creating a geodatabase for GIS information. An example of a corresponding level 2 process to Build Geodatabase would be Enter Data. This level 2 or subprocess would then deal with the data entry phase of the Build Geodatabase process. Building a model of the procedures will provide a planning tool that can eliminate duplicate processes and streamline company operations.

6.4 Improvising tools

Unfortunately, not every enterprise GIS or architecture project can be completed with software or programming right out of the box. There will always be situations where a client's needs simply cannot be met with a prepackaged solution. These situations will challenge the creativity of a consult's design team. This is a chance that someone on the design team should step up and showcase those improvisational skills and customize a program script, software package, or other tool to solve the client's problem. Every once in a while, it will actually work out that a person on your team can quickly knock out a custom solution. Online resources for open source program scripts, free small applications, developer meet ups, are great sources for a tool that can solve a small problem. Improvising tools requires an open mind, a willingness to try something new, don't be afraid to ask the online community for help or collaboration. You never know what tool you can find or create that will save time and money.

6.5 Final comment

This book has only scratched the surface of tools that are available to help create enterprise systems and GIS. Chapter 6 is just an introduction to a number of enterprise architecture tools that can easily be used for enterprise GIS. When working in the field each person will develop their own box of tools. Some of these tools will be favorites and used quite often, while some become specialty tools only used for very specific reasons or solving difficult problems.

Do not be afraid to borrow ideas for tools from other disciplines. One data management tool can be used on any other piece of data, don't let anyone ever tell you that your data is unmanageable because it was not created in their software package. Remember at the end of the day, everyone deals with data.

6.6 Summary

This chapter introduced a toolbox of standard tools that are available to anyone designing an EA/enterprise GIS. The "tools" covered are Business on a Page, business capability model, the National Association of State Chief Information Officers Maturity Model, Heat Maps, Workflow models, UML activity diagrams, along with the strategic analysis methods of Strengths, Weaknesses, Opportunities, and Threats.

A section about "advanced" tools will introduce process modeling as it applies to enterprise GIS functions. The advanced section will also expand upon the concept for current and future state modeling. Examples from NASA, conventions, and the City of Strongsville will demonstrate the backcasting technique. Backcasting is one of the best tools for creating project timelines.

Bibliography

Baer, D. 2014. "Enterprise Architecture EAOCE." *Enterprise Architecture Center of Excellence*. November 14. https://architecturescoe.org/.

Bedford, D.P.D. 2014a. *Business Architecture*. Kent, OH: Bedford, Denise Ph. D.

Bedford, D.P.D. 2014b. *Business on a Page (BOAP)*. Kent, OH: Kent State University, Dr. Denise Bedford.

Dueker, Kenneth J., J. Allison Butler. 1997. *GIS-T Enterprise Data Model with Suggested Implementation Choices*. Portland, OR: Center for Urban Studies, School of Urban and Public Affairs, Portland State University, October 1.

ESRI. 2007a. *Enterprise GIS for Local Government*. Redlands, CA, December. http://www.esri.com.

ESRI. 2007b. *GIS Best Practices Enterprise GIS*. Edited by ESRI. Redlands, CA: ESRI PRESS, January 01. http://www.esri.com.

ESRI. 2015. *Enterprise GIS. GIS Dictionary*. St. Louis, MO: ESRI, Environmental Systems Research Institute.

Gartner, Inc. 2008. "Gartner Clarifies the Definition of the Term 'Enterprise Architecture'." *Gartner Research* (Gartner) 15. doi:G00156559.

Gill, Asif. 2013. "Defining a Facility Architecture within the Agile Enterprise Architecture Context." *Orbus Software*. Orbus Software. October 1. doi:WP0107.

Graham, Andy. 2012. *The Enterprise Data Model: A Framework for Enterprise Data Architecture*. San Bernardino, CA: Koios Associates Ltd.

Joshi, Prasanna. 2016. Lecture Data Architecture. *Data Reference Cards*. Kent, OH: Kent State University, Parasanna Joshi, February 18.

Labuschagne, Louw. 2011. "Building Enterprise Architectures for Non-Architects." *Orbus Software*. September 1. doi:WP0011.

Lapkin, Anne, Philip Allega, Brian Burke, Betsy Burton, R. Scott Bittler, Robert A. Handler, Greta James. 2008. "Gartner Clarifies the Definition of the Term 'Enterprise Architect'." *Gartner*, August 12: 1–5.

Michigan Department of Information Technology. 2007. *From Vision To Action: Enterprise Architecture – Strategic Approach*. Michigan Department of Information Technology.

Ross, J., P. Weill, D. Robertson. 2006. *Enterprise Architecture as Strategy: Creating a Foundation for Business Execution*. Boston, MA: Harvard Business School Press.

Rouse, Margaret. 2007. "Enterprise Architecture (EA) Definition." *TechTarget*. June 01. Retrieved October 2015. http://searchcio.techtarget.com.

Schedlbaure, M. 2016. "Workflow Modeling with UML Activity Diagrams." *BAtimes for Business Analysts*. February 7. Retrieved February 7, 2016. http://www.batimes.com/articles/workflow-modeling-with-uml-activity-diamgrams.html.

Schroede, K., R. Tilley. 2016. "Backcasting." *Design Research Techniques*. February 12. http://designresearchtechniques.com/casestudies/backcasting/.

Schulmeisters, Kar. 2014. "Enterprise Architecture: Bridging Entrepreneurs and Hard Problems." *Orbus Software*. November 1. doi:WP0168.

Sieber, R.E. 2000. "GIS Implementation in the Grassroots." *URSIA Journal* (Vol. 12, No. 1): 15–29.

The Open Group. 19952015. *Enterprise*. January 1. www.opengroup.org/subjectareas/enterprise.

Thomas, Christopher. 2009. "What's Your Definition? Looking at What Enterprise GIS Really Means." *ArcUser* 3032. http://www.esri.com.

United States Geographic Service (USGS). 2016. *Data Standards*. http://ww.usgs.gov/dtamanagement/plan/datastandards.php.

Wilson, North Carolina. 2012. *A GIS Manager's Dialogue: Is Your Business Semi-Integrated or Fully System Integrated?* NC.

7
Designing an Enterprise GIS with EA Tools

7.1 Designing the enterprise GIS with EA tools

Chapter 7 introduces a methodology that the author developed for designing and implementing an enterprise GIS built with an enterprise archteicture (EA) foundation. The methodology is valid, and the resulting enterprise GIS has become a valuable tool for the Cities of Avon, Strongsville, and Brunswick, Ohio. This method borrows "tools" or techniques from the enterprise architecture's toolset and applies them to an enterprise GIS. This chapter demonstrates how enterprise GIS developers, designers, and programmers can bridge the gap between themselves and enterprise architecture. It is the responsibility of all GIS practitioners to use whatever tools are at their disposal to create well-organized, well-designed, and efficient enterprise systems.

The goal of an enterprise GIS is to provide the capabilities and tools of GIS to everyone in the organization. This book has discussed that allowing everyone access to GIS does create an enterprise system, but it lacks direction and cohesion and speaks with a thousand different voices instead of the clarity of one voice. The goal is to bring this system into focus so that it speaks with one voice.

Combining many voices into one will be done by using enterprise architecture as a focus to bring clarity to enterprise GIS. There are several areas where the ideas and methods of enterprise architecture fit into the GIS. A foundation of data principles, standards, and modeling creates superior GIS datasets. GISs will benefit from modeling growth, maturity, and technology. These improvements will help, but what will move enterprise GIS forward is the development of goals.

7.2 Enterprise GIS goals

Establishing goals will be the first step in providing a strategic vision of an enterprise GIS. Goal statements should clearly state the purpose of each goal. Once the goals have been determined and listed by priority, the system now has a direction for growth. New technologies, data, or opportunities can be measured against the goals and determined to be something that should be pursued or avoided.

Figure 7.1 is an example of a Goals Chart for a GIS department. Goals have a bold typeface, and their related sub-goals or "steps-to-completion" listed below (Bedford 2014). Each goal's hierarchy identification (ranking) number populates the far-left column. Each sub-goal ID number ends with a ".0" to show their relation to one of the main goals. A Goals Chart provides a summary of what will need to be accomplished to achieve the department's goals.

The chart has the columns entitled Description of Achievement, Intended Results, Measurements, and Timeframe, to provide additional information about each goal or sub-goal. The description of achievement contains an explanation that explains each goal's objective. The Intended Results include information on what change or impact has occurred on the system upon completing the goal. The column named Measurements defines a method on how to determine a goal's progress, failure, or achievement. Finally, Timeframe tells how much time it will take to complete the goal. These additional columns provide management with a more rounded explanation about what, why, how, and when the goal will be accomplished.

A list of goals can be used to follow the enterprise system's growth. Let's assume that the goal "Achieve 15 GIS users" has been listed on the goal sheet (Figure 7.2). Achieve 15 GIS users can be broken down into two or three progressive steps: "Achieve five new users by the end of year 1, five new users by the end of year 2, and five new users by the end of year 3" for a net growth of 15 users. A spreadsheet program is a tool for creating a chart for system maturity/growth.

The sample growth table demonstrates five people working with GIS in 2016. A goal of 20 users can be reached by the year 2019. Progress can be tracked by showing that five new users will be added to the system each year from 2017 to 2019. A simple shading of the table cells will create a bar graph that gives a quick and easy graphic to show the growth of the system.

The shaded areas are for demonstrating completed goals. Completed goals show that growth is continuing as projected and how much will need to be completed to achieve the final goal. These tables can be adjusted as required if growth proceeds slower or faster than what was initially projected by the design team. A Maturity Table is an excellent tool for determining when to revisit and reset the goals. Once the majority of goals have been accomplished new goals must be set to maintain and encourage future growth.

7.3 The geodatabase

A foundation stone for both enterprise architecture and GIS is the creating of a central warehouse of data and information. This concept boils down to

Designing an Enterprise GIS with EA Tools

CHAGRIN VALLEY ENGINEERING GIS DIVISION GOALS

Vision	HIERARCHY	GOAL NAME	Description of Achievement	Intended Results	Measurements	Timeframe
Company is stable with positive growth.	1.0	Business Stability	Eliminate wild swings between the ups and downs of a business life cycle.	steady progress forward	steady growth/profits	5 years
	1.1	Low turn over	control turnover of clients and employees	retention of clients and employees	number of clients and employees lost	5 years
Produces a yearly profit.	2.0	Business Profitability	show increase in profits	Establish a profitable business	income over expenses	5 years
	2.1	control expenses	eliminate unnecessary expenses	increase profits by controlling expenses	increase in earnings	5 years
Enjoys steady growth.	3.0	Business Growth	Thru Clients & Products	Grow the company	how many clients & contracts per year	5 years
	3.1	Clients	Add 1 major client per quarter	Gain new clients & prevent loss	Addition of new clients each year	5 years
	3.2	Contracts	sign 2 major contracts per year	Gain Growth thru work	Number of major contracts per year	5 years
	3.3	Recurring work	Establish recurring or renewing contracts (bread & butter work)	Steady & recurring work	Number of contracts per month	5 years
Provides a range of products.	4.0	Products	applications and programs	Establish a foundation of products		5 Years
	4.1	games	mobile, desktops, online	income	1 game per year	5 Years
	4.2	applications	develop applications	income	1 application per year	5 years
	4.3	Websites	develop turnkey websites	income	1 turnkey site per month	5 years
	4.4	Mobile	Develop apps for mobile phones	income	2 applications per year	5 years
	4.5	GIS	Develop apps and programs for mapping	income	1 map app per year	5 years
Provides contract services.	5.0	Services	Consulting, IT, & GIS	Establish a range of services	Through new contracts	5 years
	5.1	GIS Consulting	Contract analysis, programming, system design	income	1 new contact every 6 months	5 years
	5.2	IT Consulting	Contract work for public & private sector	income	1 new contact every 3 months	5 years
	5.3	Information consulting	Create custom databases, information networks, data cleaning	income	3 new contacts every 6 months	5 years
A high level of satisfaction will keep customers and employees.	6.0	Satisfaction	Customer, Owner, Employee	Retention	How many repeats or employee loss	5 years
	6.1	Owner	Satisfied with company production, services, and performances.	Retention	Client & Employee comments	5 years
	6.2	Customer	Satisfied with the company's work, services, and deliverables	Retention/Recommendations	Client comments	5 years
	6.3	Employees	Satisfied with work environment, benefits, and wages	Retention/Recruiting		
	6.4	Quality	Maintain the highest level possible.	Best Product Possible	Positive/Negative comments	5 years

FIGURE 7.1
Example of enterprise GIS goals chart. (Author's project files.)

GOAL MATURITY/GROWTH CHART				
Goal	Present 2016	Year 1 (2017)	Year 2 (2018)	Year 3 (2019)
15 GIS Users	5	10	15	20

FIGURE 7.2
Example of maturity table.

creating a folder or database that is accessible for everyone on the system. The GIS will use a geodatabase for the data warehouse. A geodatabase will locate all the available data into one container while providing the capabilities of distribution and versioning. The geodatabase will add more flexibility and adaptability to the enterprise GIS (ESRI 2007b).

The geodatabase has two main structures: feature datasets and classes. A feature dataset is similar to a windows file folder because it is used to house a group of feature classes that share a common theme. Feature classes are collections of related items which are a point, line, polygon, or text. Consider this example. A geodatabase for the City of Strongsville contains the following feature datasets: LAND, TRANSPORTATION, GENERAL. Data about the city's streets, railways, and pavement conditions are in separate feature classes that are grouped and stored in the feature dataset TRANSPORTATION. A feature dataset size is limited to 1TB, which would seem like a limitation until one realizes that a geodatabase can contain an unlimited number of feature datasets. Since there are no limits on the number of datasets, this means that a geodatabase can grow to almost any size without sacrificing performance!

Since the geodatabase can contain an unlimited amount of data, it has the potential to hold the same amount of data errors! Data standards can be used to prevent or reduce these data problems. Data standards about folder and file naming standards when applied to the feature datasets and classes allow users to identify the following quickly: the where, what, and when of data. Now a simple glance at the database will let the viewer see what is available and what is the age of the information.

Once the geodatabase's domains, feature datasets, and classes are established, it is ready to begin receiving data. A geodatabase is typically loaded with legacy data from the traditional GIS before new data is uploaded. Legacy data will have different sources, attributes, and map projections, or might be data and kept only for reference purposes. Information can be sorted, organized, tracked, and appended with new information in a Data Catalog.

Feature classes can only hold one type of data shape (point, line, or polygon). All the information about that particle article is in an attribute table within the feature. These attribute tables consist of two or more fields (OBJECTID, HOUSE NUMBER, STREET, etc.) that hold data about the items in the class. Establishing and enforcing data format standards will prevent data entry errors from entering the table. Data formats and standards are the first line of defense for the geodatabase.

Databases commonly leverage a "look up" table to help provide information or fill information in another table, particularly for a data entry application on a mobile device. These tables are pre-loaded with data values that can be used to validate entry values; if the data entered is part of the list, then the entry is accepted. If the entered data does not match a value in the table, then the entry is rejected, and the user is requested to make another choice. Leveraging "look up" tables is an excellent technique for preventing the entering of bad data.

The geodatabase can create and store these "look up" tables and make them available to any feature class table. These are called geodatabase domains (ESRI 2007) and are very useful to define and enforce data format standards. Domains can be set to one of several data formats (string, double, float, large integer, etc.) and consist of a code and description column. The code column is where the code or abbreviation representing the data, and the description defines the code value. An example would be the domain PIPE TYPE. PIPE TYPE is the value of the text, and the CODE value contains an abbreviation for pipe material, and the DESCRIPTION is the full word (Table 7.1).

Geodatabase domains are easily updated or created with MS Excel tables and the "table to domain" tool. This tool creates a domain from any MS Excel table with a two-column format, defined as CODE and DESCRIPTION no matter how many values it contains. MS Excel tools can remove data errors: typos, spelling, or proper format (i.e., date = M/D/YEAR) before the data uploaded into the domain. Domains are a tool that should be used to prevent and protect the geodatabase from data entry or format errors.

Once the geodatabase's domains, feature datasets, and classes have been established, it is ready to begin receiving data. A geodatabase is usually loaded with legacy data from the traditional GIS before new data is uploaded. Legacy data will have different sources, attributes, and map projections, or might be data and kept only for reference purposes. Information can be sorted, organized, tracked, and appended with new information in a Data Catalog.

TABLE 7.1

Geodatabase Domain for Pipe Material

CODE	DESCRIPTION
PVC	Polyvinyl chloride
CONC	Concrete
RCONC	Reinforced concrete
CI	Cast iron

7.4 Catalogs

This book would like to turn the reader's attention to three very useful inventories of data, applications, and technology. Enterprise architects refer to these inventories as "catalogs" (Bedford 2014). Building one of these catalogs is labor and time intensive, and the effort pays off in the end. These are great tools for providing a "current state" of data, technology, or software within the organization. Catalogs offer a benefit to anyone planning, budgeting, or measuring the system's growth.

Why go to the trouble of creating a catalog? A catalog puts the information available right at a user's fingertips. A simple query or click on the spreadsheet tab quickly displays essential data. Place yourself in the role of the technology architect, who is asked to recommend computers for upgrades or replacement. Before developing a technology catalog, someone must physically inspect or trust employees' assessment on which piece of equipment requires replacement or upgrade. Each computer is entered into a technology catalog; anyone can look at the lifecycle column and quickly determine which machine is ready for an upgrade or replacement. The ROI from building inventories is the time saved when looking for specific information for data, technology, or applications.

Catalogs are also excellent tools for budgetary planning. The information contained in the columns of the lifecycle, age (year installed or acquired), or status can reveal when an item is approaching obsolesces. A financial planner can avoid "spikes or surprises" since the catalog can help determine when things are ready for upgrades or replacement. These replacements or upgrades are anticipated and planned within a budget, thereby avoiding the stress and confusion created when something unexpectedly fails and must be replaced immediately, thus incurring a much higher cost. Preventing unexpected budgetary surprises is another benefit of these catalogs.

7.5 Data catalogs

Data catalogs are another tool that GIS can "steal" from the data architect's toolbox. Data catalogs do not have to be complicated and are easily formatted in a spreadsheet program like MS Excel. The advantage of using Excel is that the data catalog table can then upload into the geodatabase! A GIS data catalog at a minimum should have these fields: name, file type, data type, contents, projection, source, hyperlink (to documents, websites), and comments. Please refer to Figure 7.3 for an example of a data catalog setup.

Note that the data is divided into categories on one sheet instead of having a type on different layers of a workbook. A geodatabase will not import

Designing an Enterprise GIS with EA Tools 97

CITY KNOWLEDGE GIS DATA & LAYERS BOOK

NAME	FILE TYPE	DATA TYPE	CONTENTS	MAP PROJECTION	DATA SOURCE	SOURCE LINK	COMMENTS
SHAPEFILES							
Bndy_Cities	Shapefile	Point	Ohio Cities	NAD 1983 StatePlane Ohio South FIPS 3402 (US FEET)	Mid Ohio Regional Planning Commission	https://www.morpc.org/our-region/data-maps-tools/gis-files/index	
Census_Blocks_2010_Data	Shapefile	Polygon	12 county Area Cen Blocks	NAD 1983 StatePlane Ohio South FIPS 3402 (US FEET)	Mid Ohio Regional Planning Commission	https://www.morpc.org/our-region/data-maps-tools/gis-files/index	Data will be clipped to city boundary.
Enviro_Open_Space	Shapefile	Polygon	Parks, Cemeteries, golf courses	NAD 1983 StatePlane Ohio South FIPS 3402 (US FEET)	Mid Ohio Regional Planning Commission	https://www.morpc.org/our-region/data-maps-tools/gis-files/index	Data will be clipped to city boundary.
LBRSaddresspoints_FRA	Shapefile	Points	Addresses Franklin County	NAD 1983 StatePlane Ohio South FIPS 3402 (US FEET)	Mid Ohio Regional Planning Commission	https://www.morpc.org/our-region/data-maps-tools/gis-files/index	data source for a Geo-Coding locator
LU_GRIDS_2012MTP data	Shapefile	Points		NAD 1983 StatePlane Ohio South FIPS 3402 (US FEET)	Mid Ohio Regional Planning Commission	https://www.morpc.org/our-region/data-maps-tools/gis-files/index	
Points_of_interest	Shapefile	Points	Cultural Sites	NAD 1983 StatePlane Ohio South FIPS 3402 (US FEET)	Mid Ohio Regional Planning Commission	https://www.morpc.org/our-region/data-maps-tools/gis-files/index	Data will be clipped to city boundary.
TAZ_Tpdam12vars	Shapefile	Polygon	Traffic Analysis Zones	NAD 1983 StatePlane Ohio South FIPS 3402 (US FEET)	Mid Ohio Regional Planning Commission	https://www.morpc.org/our-region/data-maps-tools/gis-files/index	Data will be clipped to city boundary.
Trans_Bikeways	Shapefile	Polyline	Existing, Proposed, Trails	NAD 1983 StatePlane Ohio South FIPS 3402 (US FEET)	Mid Ohio Regional Planning Commission	https://www.morpc.org/our-region/data-maps-tools/gis-files/index	Data will be clipped to city boundary.
Trans_Street_Centerlines LBRS	Shapefile	Polyline	County Rd files with LBRS data	NAD 1983 StatePlane Ohio South FIPS 3402 (US FEET)	Mid Ohio Regional Planning Commission	https://www.morpc.org/our-region/data-maps-tools/gis-files/index	Data will be clipped to city boundary.
GEO-DATABASES & FEATURE CLASSES							
CorpBndry	Geo-database						
CorpBndry	Feature Class	Polygon	City of Columbus Bndry	Custom - Ohio 3402, Stourthern Zone 1983, US Survey Feet	City of Columbus	http://icolumbus.gov/Templates/Detail.aspx?id=68551	
Forestry	Geo-database						
Columbus Trees	Feature Class	Point	Tree Info	NAD 1983 StatePlane Ohio South FIPS 3402 (US FEET)	City of Columbus	http://icolumbus.gov/Templates/Detail.aspx?id=68551	
Forestry Zone	Feature Class	Polygon	Tree Zones	Custom - Ohio 3402, Stourthern Zone 1983, US Survey Feet	City of Columbus	http://icolumbus.gov/Templates/Detail.aspx?id=68551	
Park Facility	Feature Class	Point	Facilities by type/use	NAD 1983 StatePlane Ohio South FIPS 3402 (US FEET)	City of Columbus	http://icolumbus.gov/Templates/Detail.aspx?id=68551	Covers sports, parks, community centers, recreation
Public-Safety	Geo-database						
CruiserDistricts	Feature Class	Polygon	Boundary	NAD 1983 StatePlane Ohio South FIPS 3402 (US FEET)	City of Columbus	http://icolumbus.gov/Templates/Detail.aspx?id=68551	
Firestations	Feature Class	Polygon	Station Location/Type	NAD 1983 StatePlane Ohio South FIPS 3402 (US FEET)	City of Columbus	http://icolumbus.gov/Templates/Detail.aspx?id=68551	Location shown as a parcel - not address
Hospitals	Feature Class	Point	Hospital Location	NAD 1983 StatePlane Ohio South FIPS 3402 (US FEET)	City of Columbus	http://icolumbus.gov/Templates/Detail.aspx?id=68551	Location tied to an address point
PolicePrecincts	Feature Class	Polygon	Boundary	NAD 1983 StatePlane Ohio South FIPS 3402 (US FEET)	City of Columbus	http://icolumbus.gov/Templates/Detail.aspx?id=68551	
PoliceSubstations	Feature Class	Polygon	Station Location/Type	NAD 1983 StatePlane Ohio South FIPS 3402 (US FEET)	City of Columbus	http://icolumbus.gov/Templates/Detail.aspx?id=68551	Location shown as a parcel - not address

FIGURE 7.3
Example of data catalog. (Author's project files.)

a multi-sheet table, so one has to create separate tables for each category of data or create one large table. It is not necessary to load the data catalog into the geodatabase's table library.

The data catalog will keep data organized and aid in identifying potential issues with incorporating legacy data into the enterprise system format (Bedford 2014). A catalog of the data is an immense help when establishing metadata files for the system. There are numerous metadata templates available for download, and it is beyond the scope of this book to discuss how to choose one. This data catalog will simplify transferring information about a dataset into a metadata file because one can use the "copy and paste" command to transfer the information into the proper metadata field. Although not as elaborate or detailed as the premade metadata templates, the data catalog could be used as a metadata library.

The importance of the data catalog can be found in two pieces of information. The source column is essential since that reveals where the data was obtained. This allows anyone to look at the data's source and determine if it is a trustworthy source. Another column is lifecycle or status. This column predicts or reveals when the information is considered out of date. Accurate and current (or up to date) data is critical to establishing a system's credibility. A single piece of data that is out of date or inaccurate can destroy users' trust in an information system.

Regaining someone's trust in an information system after they have been burned by lousy data is very hard. A person who has been embarrassed, surprised, or humiliated after a member of the audience reveals that the presentation is inaccurate will not trust the system and probably make everyone else aware of the system's problems. All it takes is one person with an issue to destroy the entire organization's belief in the system's data. Remember the system can only provide value to the organization if people trust its information. When trust is lost, the system will fail, and it is very hard to regain confidence.

7.6 Technology catalog

The next two catalogs to be discussed are technology and software applications (Bedford 2014). These are handy tools for technology and application architects. Each architect (or manager) enjoys an excellent return on investment (ROI) for the time committed to building and maintaining the data. Once again, the person creating an enterprise GIS can also reap the benefits of setting up these two catalogs.

Let's begin this discussion with the technology catalog (inventory). This is essentially a listing of every piece of hardware used by the organization or department. One will find computers, phones, printers, etc. within the pages or sheets of a technology catalog. The purpose of this catalog is to gather information for all of the organization's technology into one easily accessible location.

Google sheets, MS Excel, or any spreadsheet program can be used to create a technology catalog. One must pay attention to the details when developing the sheet's layout that will house the technology information. Remember, the purpose of this catalog is to capture and maintain as much information as possible on any of the technology like computers, monitors, printers, etc., that is used or interacts with the enterprise GIS. The catalog breaks the hardware down into categories and shows where each piece of equipment fits into the overall system.

Examine the technology catalog displayed in Figure 7.4. Do you notice how much detailed information is recorded about computers? The column names provide a clue as to how much detail for each item is being recorded. Here is a closer look at these columns. The first column is the home system, and this is the unique identifier for each item. JWOODARD-WIN7 reveals that this computer has been assigned to JWOODARD and uses the operating system Windows 7. The description is where the issue is described. The computer assigned to JWOODARD is an HP Z201 Workstation. This information is recorded from the documentation provided by the supplier and manufacturer. Item is the column that classifies the JWOODARD-WIN7 as a desktop computer.

The next 11 columns record even more details about the desktop computer JWOODARD-WIN7. MS number is where the Microsoft serial number (00371-OEM-8992671-00008) for the operating system is recorded. This is followed by S/N#, which is the serial number assigned to the computer or internal devices by the manufacturer. P/N# is a column that records the part number for the hard drives. The column OS (operating system) defines the operating system name and service pack number. All of this information can be recorded from the manufacturer labels found on the back of the computer and secondary devices.

Once the basics are recorded, the catalog moves into more in-depth detail about the computer system. Information for the hard drives is recorded in the columns and size (7200 rpm and 1 TB). Statistics for the computer's central processing unit (CPU) is entered into the columns CPU NAME, CPU# (number of CPUs) 8(2/6v); speed of each CPU is recorded in CPU 1 (3.40 GHz), CPU 2 (3.40 GHz); RAM (16 GB) records how much memory has been installed. All of these details are arranged in a particular pattern; the computer is in the first row, while individual components (graphics cards, internal/external hard drives, etc.) are listed in the rows beneath. This arrangement is a quick snapshot of all the devices that might need to be upgraded or replaced in the computer.

The final columns contain information about the primary mission (purpose), category, user, and status of the computer. The column mission is where the item's primary goal is declared and recorded. Consider the illustration

CVE TECHNOLOGY CATALOG

HOME SYSTEM	DESCRIPTION	ITEM	MS NUMBER	S/N#	P/N#	OS	RPM	CPU NAME	CPU#	CPU 1	CPU 2	RAM
JWOODARD-WIN7	HP Z201 WORKSTATION	DESKTOP COMPUTER	00371-OEM-8992671-00008	2UA136153W		Windows 7 SP1		Core i7-2600	8(2/6v)	3.40 GHz	3.40 GHz	16 GB
JWOODARD-WIN7	NVIDIA QUADRO 4000	GRAPHICS CARD	NONE			Windows 7 SP1						
JWOODARD-WIN7	SEAGATE BARRACUDA	INTERNAL HD	NONE	9VPBDEXM	ST31000528AS	Windows 7 SP1	7200					
JWOODARD-WIN7	HITACHI	INTERNAL HD	NONE	JP2940J825795V	HDS721010CLA832	Windows 7 SP1	7200					
JWOODARD-WIN7	WESTERN DIGITAL	INTERNAL HD	NONE	WCAZA8279815	WD20EARX-00PASB0	Windows 7 SP1	7200					
WESTERN DIGITAL	WD MY BOOK	EXTERNAL HD	NONE	WCAZAH256042	WDBACW0020HBK-01	Windows 7 SP1	7200					
SEAGATE	SEAGATE	EXTERNAL HD	NONE	2GEY6FBW	9ZC2A8-501	Windows 7 SP1	7200					
TSNODE-600 PRO	HP COMPAQ 600 PRO	DESKTOP COMPUTER	00371-OEM-8992671-00008	2UA0370MXD	VS826UT#ABA	Windows 7 SP1		Pentium (R) Dual	2	3.07 GHz	3.07 GHz	8 GB
TSNODE-600 PRO	HARD DRIVE	INTERNAL HD	NONE		ST3160318AS	Windows 7 SP1	7200					
TSNODE-600 PRO	ATI RADEON HD 4650	GRAPHICS CARD	NONE			Windows 7 SP1						
CVE-HP8300-B	HP8300 - B	DESKTOP COMPUTER	00371-OEM-9309205-70166	2UA24T08PR	C9H33UT#ABA	Windows 7 SP1		Core i7-3770	2	3.4 GHz	3.4 GHz	8 GB
CVE-HP8300-B	NVIDIA QUADRO PRO 2000	GRAPHICS CARD	NONE			Windows 7 SP1	NONE					
CVE-HP8300-B	HARD DRIVE	INTERNAL HD	NONE	ST500DM002-BD142		Windows 7 SP1	7200					

SIZE	MISSION	CATEGORY	USER	STATUS
	GIS PROCESSING	PRODUCTION LAYER	JRW	OK
	GIS PROCESSING	PRODUCTION LAYER	JRW	OK
1 TB	DATA STORAGE	STORAGE LAYER	JRW	OK
1 TB	DATA STORAGE	STORAGE LAYER	JRW	OK
2 TB	DATA STORAGE	STORAGE LAYER	JRW	OK
1.5 TB	DATA STORAGE	STORAGE LAYER	JRW	OK
	GIS PROCESSING	PRODUCTION LAYER	AK	NEED MEM
140 GB	DATA STORAGE	STORAGE LAYER	AK	NEED HD
	GIS PROCESSING	PRODUCTION LAYER	AK	OK
	DATA STORAGE	PRODUCTION LAYER	PZ	NEED MEM
	GIS PROCESSING	PRODUCTION LAYER	PZ	OK
500 GB	DATA STORAGE	STORAGE LAYER	PZ	NEED HD

FIGURE 7.4
Example of tech catalog. (Author's project files.)

computer JWOODARD-WN7, and its primary purpose is GIS processing. JWOODARD-WN7 is categorized in the next column as a member of the production layer. This designation means that it is used from producing GIS maps, analysis, programming, or development. Note that a HITACHI internal hard drive mission is listed as data storage, and it is categorized as a member of the storage layer.

There are two columns left in this table: user and status. The user column records the initials of the person assigned to the computer system. The designation JWOODARD-WN7 is the original designation for the computer, and, however, if JWOODARD has left the company or received a new computer, his old machine might be reassigned to a new user. Consider the illustration's example of TSNODE-600 PRO; you would expect to see the initials TSN in USER. However, the USER column shows that AK now is using the machine, representing a reassignment of the technology.

Status is the final column of the table, although last, it is by no means the least of the columns that create the catalog. A status column is one of the most useful information tools found within the table. Why? A glance at status informs the viewer which pieces of technology are OK, which NEED something, or which should be disposed of.

There is a substantial investment in time for gathering and recording all of the data about the organization's technology. The time invested will lead to an excellent return in terms of efficiency and quickness when planning to replace or upgrade a system. Rather than replace every piece of hardware in order, the catalog status column can help to select parts of equipment for replacement or update. Efficient planning and management of the organization's technology will lead to savings in effort, manhours, and money. These are the real benefits of the technology catalog.

7.7 Application catalog

The next and last catalog to discuss is about all of the applications used within the organization or department (Bedford 2014). The application catalog is an inventory of all the applications that are used by or interact with the enterprise GIS, just like the technology catalog creating a directory for all of the software applications used in the organization will require a substantial investment in time and effort.. Small companies or departments often combine the responsibility for updating, repairing, or evaluating new technologies or applications into one person's role with the title technology/applications architect.

A properly constructed application catalog will provide many benefits to someone fulfilling the responsibilities of the application architect. Much like the technology or data catalog, Google sheets, MS Excel, or any spreadsheet program can be used to create an application catalog. Again, attention to detail is required when creating the catalog's layout. Just like the other catalogs, the purpose of this catalog is to gather information. However, the data is about software applications (Adobe Photoshop, ArcGIS, MS Word, MS Access, MS Excel, etc.) that are installed on computers. Figure 7.5 is an example of an application catalog.

Spend a few minutes examining the example of the application catalog. This layout is similar to the one used for the technology catalog. First, there is a unique ID number assigned to each piece of software (column ID). The columns Application (name), Version number, Level (skill level), and Vendor (seller) all record the necessary information about the software application. Consider the software application with the ID number 2. The name of the software is ArcReader (application), version 10.1, the required skill level is basic (Level), and the Vendor is ESRI. This covers the necessary information about ArcReader; it is in the remaining columns that the applications catalog begins to shine.

Now one can begin to record how the software fits into the organization/department's business operations. First, one must decide what business capability depends upon the software, which is recorded in the column "Associated Business Capability." Then the software is assigned a grouping or family (see the column Family) and provided with a description. Next, every other application that has received support from the application is listed in the Other App Dependency column. Locate ArcGIS (ID number 1) and note how many other programs it uses: MS Word, Excel, Access, ET Geowizards, Adobe Photoshop, and Dreamweaver.

Did you find it surprising how many different software applications can be used to support map creation in ArcGIS? An immediate benefit of this catalog is the ability to determine the interconnections between the software applications. ArcGIS is the primary GIS mapping software, yet App Dependency reveals that there are several other applications that support mapping projects. These secondary applications provide text editing, graphics, interactive programming scripts, etc. that allow a standard map to become extraordinary. Any architect examining this catalog can quickly deduce which applications are essential to support company operations or projects.

The next set of columns provide reader information about licenses, status, users, sharing, and computer(s) locations that have the applications. It is essential to track the number of users vs. available licenses. Today's software only provides a limited number of licenses per purchase, and if there are more users than licenses, someone will not be able to work. This could lead to project delays, arguments between staff, and possible loss in production. This can be avoided by planning the purchase of future licenses as production demands increase.

The last section covers employee satisfaction with the software when it might be replaced or updated, and the all-important question "who pays for it?" Take another look at ArcGIS or ID #1 in Figure 7.5. The software has

Designing an Enterprise GIS with EA Tools

CVE APPLICATION INVENTORY

ID	APPLICATION	VERSION	LEVEL	VENDOR	ASSOCIATED BUSINESS CAPABILITY	FAMILY	DESCRIPTION	OTHER APP DEPENDENCY	#LICENSES	STATUS	# USERS	SHARED	STATION LOCATION	RELIABILITY	USER SATISFACTION	COMMENTS	SCHEDULED RETIREMENT	COST CENTER
1	ArcGIS	10.1	Basic	ESRI	GIS	GIS program	SOFTWARE THAT ENABLES A USER TO CREATE DATABASES, QUERY INFORMATION, AND PERFORM INFORMATION MAPPING	MS Word, Excel, Access, ET GeoWizards, GeoTools, Adobe Photoshop, Dreamweaver	3	ACTIVE	3	NO	35, 36, 37	GOOD	POOR	FREQUENT CRASHES	10.2 SP1	GIS
2	ArcReader	10.1	Basic	ESRI	GIS	GIS program	A FREE GIS VIEWER THAT CAN READ ONLY PMF FILES FROM ARCGIS	ESRI ArcGIS	20	ACTIVE	35	YES	1,2,3,4,5,6,7,8,10,11,12,13,14,15,16,17,36,37	EXCELLENT	EXCELLENT	NO PROBLEM	10.2 SP1	GIS
3	ETGeoWizards	10	N/A	ET	GIS	GIS EDITING TOOLS	PROVIDES A SUITE OF EDITING TOOLS TO ARCGIS BASIC	ESRI ArcGIS	1	ACTIVE	1	NO	37	EXCELLENT	EXCELLENT	NO PROBLEM	10.2 SP2	GIS
4	ETGeoTools	10	N/A	ET	GIS	GIS EDITING TOOLS	PROVIDES A SUITE OF EDITING TOOLS TO ARCGIS BASIC	ESRI ArcGIS	1	ACTIVE	1	NO	37	EXCELLENT	EXCELLENT	NO PROBLEM	10.2 SP3	GIS
5	Adobe Acrobat Pro	8	PRO	Adobe	GIS / TEACHING / ADVISING / PROGRAMMING/RESEARCH & DEVELOPMENT	PDF Tool Suite	ALLOWS USER TO CREATE AND EDIT PDFS. ALSO ALLOWS CREATION OF MULTI-FILE PDFS	ANY FILE TO BE CONVERTED TO PDF	4	ACTIVE	4	NO	17,35,36,37	GOOD	SATISFIED	HARD TO LEARN	WILL RENEW WHEN NO LONGER SUPPORTED BY OS	GRAPHICS
6	Adobe Photoshop	4	N/A	Adobe	GIS / TEACHING / ADVISING / PROGRAMMING/RESEARCH & DEVELOPMENT	GRAPHIC DESIGN	SOFTWARE THAT ALLOWS USER TO EDIT AND MANIPULATE DIGITAL IMAGES	NONE	1	ACTIVE	1	NO	37	GOOD	SATISFIED	HARD TO LEARN	WILL RENEW WHEN NO LONGER SUPPORTED BY OS	GRAPHICS
7	Adobe Dreamweaver	4	N/A	Adobe	GIS / TEACHING / ADVISING / PROGRAMMING/RESEARCH & DEVELOPMENT	GRAPHIC DESIGN	SOFTWARE THAT ALLOWS USER TO CREATE, EDIT, UPDATE WEBSITES OR PAGES	MS Word, Excel, Access, Adobe Photoshop, Fireworks	1	ACTIVE	1	NO	37	GOOD	GOOD	HARD TO LEARN	WILL RENEW WHEN NO LONGER SUPPORTED BY OS	GRAPHICS
8	Adobe Fireworks	4	N/A	Adobe	GIS / TEACHING / ADVISING / PROGRAMMING/RESEARCH & DEVELOPMENT	GRAPHIC DESIGN	SOFTWARE THAT ALLOWS USER TO CREATE IMAGES OR GRAPHICS FOR WEBSITES AND WEB PAGES	MS Word, Excel, Access, Adobe Photoshop, Dreamweaver	1	ACTIVE	1	NO	37	GOOD	GOOD	HARD TO LEARN	WILL RENEW WHEN NO LONGER SUPPORTED BY OS	GRAPHICS
9	MicroSoft Word	2007	N/A	MicroSoft	GIS / TEACHING / ADVISING / PROGRAMMING/RESEARCH & DEVELOPMENT	OFFICE	SOFTWARE THAT ALLOWS USER TO CREATE, EDIT WORD DOCUMENTS FOR CLIENTS, BUSINESS OR GIS.	NONE	35	ACTIVE	35	NO	ALL	EXCELLENT	EXCELLENT	NO PROBLEM	WILL RENEW WHEN NO LONGER SUPPORTED BY OS	OFFICE
10	MicroSoft Excel	2007	N/A	MicroSoft	GIS / TEACHING / ADVISING / PROGRAMMING/RESEARCH & DEVELOPMENT	OFFICE	SOFTWARE THAT ALLOWS USER TO CREATE, EDIT SPREADSHEETS OF DATA FOR THE BUSINESS OR GIS	NONE	35	ACTIVE	35	NO	ALL	EXCELLENT	EXCELLENT	NO PROBLEM	WILL RENEW WHEN NO LONGER SUPPORTED BY OS	OFFICE
11	MicroSoft Powerpoint	2007	N/A	MicroSoft	GIS / TEACHING / ADVISING / PROGRAMMING/RESEARCH & DEVELOPMENT	OFFICE	SOFTWARE THAT ALLOWS USER TO CREATE, EDIT PRESENTATIONS FOR CLIENTS OR THE BUSINESS	NONE	10	ACTIVE	10	NO	1,3,5,6,14,15,18,35,36,37	EXCELLENT	EXCELLENT	NO PROBLEM	WILL RENEW WHEN NO LONGER SUPPORTED BY OS	OFFICE

FIGURE 7.5
Example of applications catalog. (Author's project files.)

been giving a reliability rating of "Good," not "Excellent," and user satisfaction is showing "Poor." This seems to be an odd rating for a software package that is considered an industry standard. When the comments column is examined, a reader will find the reason for the various grades, and users are reporting "frequent crashes." This information should prompt a closer look at the software.

ArcGIS is not cheap to replace! However, it does seem to be a problem for the organization. A quick look at the software's scheduled retirement shows that the current 10.1 version will be replaced when 10.2 SP1 is available. The company can wait for the new version or possibly pay for in-depth tech support to fix the problem. Cost center is the column that shows what department, person, or company will be financially responsible for this decision. A quick look and one finds that the GIS department will be held accountable for this software fix.

Information in the data, technology, and application catalogs is not mutually exclusive. Instead, they provide the most significant planning budgets when used together. Let's consider this scenario: a new software application is being purchased. The application catalog allows planners to determine how the new software interacts with existing programs and projects. A quick check of the technology catalog would reveal which computers' current hardware would or would not be able to support the software. Now, if a hardware upgrade is necessary, a machine that is closest to the software's requirements can be selected for a minor update to meet the new specifications.

Using the information in technology and applications, catalogs have saved the organization time and money. A time savings occurred since planners did not have to use a trial period to determine if the new software would successfully interact with established applications.

The cost (money) savings were realized since a minor upgrade would allow a computer to support the new software, and the expensive purchase of a new machine was avoided. Building these catalogs requires an investment in time and effort. This investment will generate a high ROI by realizing savings in cost, time, and resources.

7.8 Current state to future state model

Enterprise architects or system designers have two tools that help them evaluate an existing system, organization, or department and then determine how it should exist in a future version. These tools are known as the "Current and Future State" documents. The "current state" is an evaluation of the organization and assessment of where it stands in the present. Establishing a good understanding of the current state of the organization is the first step in determining a plan for the future (Bedford 2014).

Once the current state of the organization has been documented and understood, the next step is to create a future vision (state) of the organization. This future vision of the organization could be 5, 10, or even 15 years from the current state. A future state document has two functions: it determines a future vision of the company and is the goal that will be fulfilled by following the enterprise architecture or plan. Remember that when the future state has been achieved, it becomes the current state, and a new future vision is established. This begins a cycle of current state to a future state to the current state, etc. that continually pushes the organization forward while improving it (Bedford 2014).

The foundation for completing a current state to future state evaluation has already been laid! Just look to the goal statements, data, technology, application catalogs, and the BOAP. Each of these items will contribute toward establishing a good understanding of the current state of the organization. Figure 7.6 demonstrates how a current state to future state table is designed. Goals are used to load information into the fields: Goal Contributes to, Goal Hierarchy ID, and Goal Name. Description of achievement is a restatement of the goal's definition, and Intended Results is a statement of what is expected to happen within the system once the goal is attained. The field Measurement is important because this shows what device or achievement will measure if the goal has been successfully achieved.

The last three columns on the right are labeled current state (Year), midpoint milestones (Year), and future state (Year). Current state records what the present status for each goal is. This also marks the start of the planning period, if a five-year plan is being designed, then the year on current State will be 2016, and the future state will be 2021. The midpoint milestones will be given a year halfway between 2016 and 2021, which is 2.5 years or which would mathematically be 2018.5, which when rounded down is 2018 A milestone marker is demonstrating a mid-step between the starting and final attainment of the goal (Bedford 2014a).

The future state is listed in the final column because achieving this state marks the end of planning, work, and end of a cycle from the current state to future state. The future state can only be achieved when the midpoint milestones have been completed. This does not mean that everyone should get a pat on the back, relax, and let the business run on cruise control. When the future state has been achieved, it becomes the current state and is the beginning of a new planning cycle to reach the next future state.

This process can be demonstrated with the help of Figure 7.6. Consider the first line of the current state to future state model. Reading from left to right, the table provides a breakdown of goal 1.0. The goal is a GIS and is described as "create an efficient (GIS)." The intended result is "A GIS that responds quickly to the needs of CVE or clients." The success of this goal will be determined by "the amount of positive feedback" received from clients and staff.

The last three columns will establish the current state for each goal, the midpoint milestone, and the future state. Note that this table has a five-year

CVE GIS DIVISON: CURRENT STATE TO FUTURE STATE

Goal Contributes To (Vision)	Goal Hierarchy ID	Goal Name	Description of Achievement	Intended Results	Measurements	CURRENT STATE YEAR: 2014	MIDPOINT MILESTONES YEAR: 2016.5 (2.5 YEARS)	FUTURE STATE YEAR: 2019 (5 YEARS)
GIS SYSTEM	1.0	GIS SYSTEM	Create an efficient GIS	A GIS that responds quickly to the needs of CVE or clients.	The amount of positive feedback	SYSTEM DESIGN UNDERWAY / SOFTWARE SELECTED & INSTALLED	SYSTEM ESTABLISHED AND SUPPORTS 65% OF ALL COMPANY PROJECTS	SYSTEM ESTABLISHED AND SUPPORTS 95% OF ALL COMPANY PROJECTS
1.0	1.1	Eliminate System Redundancies	Eliminate, Remove, Prevent Redundancies occurring in the GIS system.	A streamlined mapping and data flow	Time saved/number of duplicate files/Employee feedback	GIS system Beta version implemented and data flow mapping started.	GIS system with a data architecture is established. Duplicate files reduced to 25% of total file count.	GIS enterprise architecture and system established. Duplicate files reduced to 5% of total file count.
1.0	1.2	User Friendly	Make the system quick and easy to understand	Create a system that is easy to learn and use.	User feedback/number of users/	GIS Viewer installed on 7 CVE & 4 municipalities' computers. Positive feedback rating 25%	GIS Viewer installed on 14 CVE & 12 Municipalities' computers. Positive feedback rating 60%	GIS Viewer installed on 14 CVE & 12 Municipalities' computers. Positive feedback rating 95%
1.0	1.3	Keep Technology updated	Keep system hardware & tech current	Prevent system response lag times. Prevent "rush" or reactive upgrades to software & technology.	Technology & hardware age by year. Hardware should be 5 years from Market. Software Applications Should be on 2 versions older than current.	Computer Hardware purchased 2012. ArcGIS software version 10.2. Graphics, Office, Misc software are 1 version off from current market.	All hardware purchased at start of 2016. ArcGIS software upgraded to current version. New versions of the Graphics, Office & Misc. software purchased in 2016.	Computer Memory, Hardwares upgraded to current version. New versions of the Graphics, Office & Misc. software purchased in 2019.
ACCURATE DATA	2.0	ACCURATE DATA	Eliminate inaccuracies in the GIS information	To create a accurate dataset that is trusted by GIS users.	Decrease in errors/Decrease number of data "fixes"/Decrease complaints about dataset errors	GIS information has been obtained and rudimentary Geo-Databases created.	Importation and creation of data policies established that result in a 5% margin of data error. 30% comments are complaints.	Importation and creation of data policies established that result in a 1% margin of data error. 2% of comments are complaints.
2.0	2.1	Clean data	Develop a data "scrubbing" model	Automate basic data cleaning methods	Decrease the time & manpower required to clean datasets/Increase the accuracy of the dataset.	"Data scrubbing" processes are being established. No automation at this time. FTE hours: 120	"Data scrubbing" processes are established. First totally Automated data processing model in use. FTE hours: 60	"Data scrubbing" processes are fully automated & require minimal staff oversight. FTE hours: 8
2.0	2.2	data/information standards	Develop data/information standards about how to use, import, collect or distribute the datasets.	Maintain a consistent methodology for working with company data. Remove confusion about what can be done with the information.	Reduce the number of data "misuse or released" incidents" over the course of a business quarter.	Examine data/information standards, that are in use by other companies, municipalities. Establish the first draft of a company wide set of standards.	GIS Data Standards. Principles and process are established. Accidental Data Incidents drop from a rating of common to uncommon.	GIS Data Standards. Principles and process are refined. Accidental Data Incidents drop from a rating of uncommon to rare.
2.0	2.3	Data Security	Develop standards, methods, and procedures for protecting company & client sensitive data.	Prevent the misuse, loss or theft of company or client data/information.	Reduce the number of loss, accidents or theft incidents over the course of a business year.	Work with IT staff to integrate GIS practices with current standards, methods, and procedures for protecting sensitive data.	Safeguarding information follows established standards, methodologies, and procedures resulting no loss information. IT & GIS committee for security policies established	Safeguarding information follows established standards, methodologies, and procedures resulting no loss information.

FIGURE 7.6
Current state to future model.

planning cycle from 2014 to 2019. Midpoint milestones will be evaluated at the two-and-half-year mark which would mathematically be 2016.5 which when rounded down is 2016.

The current state of the goal GIS is "System design underway, the software selected, and installed." Year 2016 the company should accomplish the midpoint milestone of "system establishes and supports 65% of all company projects." 2019 is the ending year for this planning cycle and if the "system is established and supports 95% of all company projects" one can say that the goal of the GIS has fulfilled the future state. The organization will be considered in the future state when 90–95% of its stated objectives have been accomplished.

Midpoint milestones are significant to the success of this model. A midpoint evaluation can reveal if the organization is on pace to succeed in moving the goals into a future state. We are ensuring that each midpoint has a measurable goal. Percentages are a quantifiable indicator, the current state could be 0%, and then the midpoint marker could be 25% improvement. A goal "staff use the system," and a midpoint marker that "even more staff use the system" are not reasonable goals. Adding percentages help measure the goals. Current state "0% of the staff use the system," and the midpoint marker would be "25% of the staff use the system." Finally, the future state might be "65% of the staff use the system." The percentages demonstrate the progress toward accomplishing the future state.

Current state to future state models is a handy tool for system designers and users. They provide a quick explanation and a way to ensure that the organization system is progressing. Everyone should remember that the system is neither static nor written in stone. A few goals will be accomplished earlier than expected, progress on other goals will be as expected, and progress on one or two goals might be incredibly slow. Feel free to make adjustments to the model to account for these issues. The trip from the current state to the future state is like a ship's journey across the ocean. Account for the progress of goals with minor course adjustments, while always keeping the ship's bow toward the port of future state, and you will have a successful voyage.

This text has introduced many new concepts, methodologies, and models for developing an enterprise system. What happens when a new system is deployed? How does someone determine if the system is growing or failing? A current state to future state model does measure progress, but it falls a little short of the mark for tracking system maturity. Fortunately, a simple model can be created to allow designers to measure how the system matures.

An enterprise GIS maturity model will measure the organization's progress in accepting, implementing, and deploying the system. A maturity model has the advantage of demonstrating how different areas are improved from the deployment of the enterprise system. There are four sections in the maturity model illustrated in Figure 7.7: management, GIS planning, GIS framework, and GIS integration. The maturity model will measure the

CHAGRIN VALLEY ENGINEERING GIS MATURITY MODEL

GIS (NASCIO)	LEVEL 0 (NOTHING)	LEVEL 1 (CREATING AWARENESS)	LEVEL 2 (ESTABLISHING A FOUNDATION)	LEVEL 3 (DEVELOPMENT)	LEVEL 4 (COMPLETION)	LEVEL 5 (EXPANSION)	LEVEL 6 (OPTIMIZATION)
Management	There is no one delivering a coordinated effort for the CVE GIS EA.	1. Begin to create the idea that the company needs a GIS EA architecture to deliver support to company Data/information mapping, creation, use, and distribution. 2. Educate owners/management/staff on the potential benefits of GIS architecture	Demonstrate that GIS is being used by CVE as a reactive procedure. Several people use GIS in Ad Hoc way to produce quick maps or illustrations. Use several small projects to demonstrate to owners/ executives, CAD Techs, & Engineers how GIS can support their projects.	Owners/ executive management aware of potential benefits of a GIS architecture. Fully support the hiring/appointment of a GIS Project Manager and create a GIS department. Begin requiring GIS to be consulted on new data or mapping projects.	Owners/ executive management accept and fully support the GIS architecture and GIS Manager. Staff begins to access and incorporate GIS into their daily routines and projects.	Owners/ executive management allocate a budget for training, hiring staff, software and tech updates. GIS is considered and accessed at the first stages of a project. Not brought in as an afterthought.	Owners/ executive management allocate a budget for training, hiring staff, software and tech updates. GIS is considered and accessed at the first stages of a project. Not brought in as an afterthought.
GIS Planning	No plan in place	Note the lack of organization, sudden shutdowns, unexpected data loss, disagreements over what to do and how to proceed.	Meet with owners and document profit loss due to duplication of software, procedures, data, and employee efforts. Work with a committee to identify company needs.	Meet with Partners, Jr Partners, & supervisors to learn about their expectations for a GIS Architecture. Begin establishing goals, standards, & procedures.	GIS Architecture is implemented. GIS staff, CVE engineers, & clients accessing data/information. Architecture has goals, standards, procedures, & a Review Board	GIS Architecture is expanded to include other departments and remote clients. Begins to shift into a true Enterprise Architecture for CVE departments and sister company.	GIS Architecture has grown into a fully functional enterprise system for all 14 client cities. MBIS (sister company). The architecture is overseen by a team of designers who report to the business & enterprise architect.
GIS Framework	There is no framework or support for the GIS architecture	Note the lack of an organized methodology to prevent confusion among employees and duplication of project efforts. Begin meeting with supervisors & partners to open lines of communication between projects.	Create the initial outline of procedures for displaying, sharing and incorporating GIS into engineering projects & CAD drawings. Begin laying out a chart for all GIS capabilities and what could be currently supported	Implement the framework as part of the GIS architecture, map-out software applications, map-out what projects are being supported by which datasets.	Define the current framework and finish a complete map with inventory. Plan the future state of the framework.	The GIS Framework complements and supports the Enterprise architecture. New models and capabilities added to the system.	GIS framework is trusted and used throughout the company. It is considered a good template framework to follow when establishing GIS systems for client businesses and cities.
GIS Integration	GIS Stand alone machine & projects. No integration.	GIS projects, data are isolated and not supporting CVE Engineering Projects.	GIS starts supporting wetland delineation projects. GIS support requested for highway and pavement maintenance projects.	CVE staff & partners trust GIS to provide quality maps, analysis for projects and grant proposals.	GIS support is expected and solicited for all projects. GIS is considered a critical member of project teams.	GIS is requested to represent CVE at client conferences and seminars. GIS helps to sell services for CVE.	GIS actively brings in new projects and revenue for the company. It is considered a valuable department that is involved in all CVE projects and processes

FIGURE 7.7
Maturity growth chart. (Author's project files.)

progress of each unit as they work toward developing a fully functional enterprise system.

This text uses maturity models (Figure 7.7) based on the National Association of State Chief Information Officers (NASCIO) framework. The NASCIO framework defines six levels of maturity. These levels are 0 – nothing, 1 – creating awareness, 2 – establishing a foundation, 3 – development, 4 – completion, 5 – expansion, 6 – optimization, for reviewing a system's maturity development (NASCIO 2015). Levels 0 through 6 are the various stages of maturity that the system must pass through before it reaches full potential. Each of these levels is given a small description, and if this is done in MS Excel, then when the level is surpassed, it can be filled in with color. This allows the model to function as a quick and easy bar graph demonstrating where each level of the enterprise architecture is within the model.

The four aforementioned sections of the company have been placed into the CVE GIS (NASCIO) column. These sections have been identified as critical items that will contribute to the system's success or failure. Management must "buy-in," accept, and most importantly, visibly support the enterprise system. Unfortunately, it is relatively common for company management to attempt to build a GIS, without understanding or comprehending the field, technology, or methodologies. Enterprise GIS designers must teach management and employees about the benefits of working with the new system. This model will enable system designers to estimate the effectiveness of the training and explanation cycle. Once management understands the benefits, cost savings, and see an upward tick in profits, they will fully support the new system initiative.

Let us examine how the maturity model will work in measuring the organization's progress with GIS planning. Currently, GIS planning is at level 0 – nothing; there is no plan in place. The company has hired a GIS professional who immediately begins logging the following system issues – disorganization, sudden shutdowns, data loss, and staff disagreements – and begins to formulate a response. This moves the GIS planning into level 1 (creating awareness); the professional documentation of these issues and reporting to management has just created awareness.

Now management desires a meeting to understand the situation. Our GIS professional demonstrates how the duplication of software, procedures, system problems, and employee confusion have created losses in time and money. The response is to form a committee that identifies issues and needs and makes recommendations to management. A foundation of planning to solve the problems with an enterprise GIS moves GIS planning into level 2 – establishing a foundation.

Meetings between executive management, department heads, supervisor, and the enterprise architect or his team to discuss a GIS architecture. These meetings will determine everyone's expectations and needs for the enterprise GIS. The enterprise architect, along with the design team works with employees to establish business goals, standards, and procedures.

This activity shows that GIS planning has reached level three – the development of a system.

Between levels 3 and 4, a series of events; installation of software, hardware upgrades, acceptance and implementation of GIS standards, procedures, and project methodologies will be completed. When the enterprise architecture has been implemented, employees and clients are benefiting from GIS analysis, mapping, data, and information; a successful planning cycle has been completed. The section for GIS planning can be moved into maturity level 4 – complete.

Levels 5 and 6 are the last two remaining columns on the maturity chart. Level 5 is about the expansion of the newly completed system into other company departments and expanding GIS capabilities through new hires, technology, and software. It also marks the transition from a small enterprise system into a larger one that encompasses remote sites. Reaching level 5 is a fantastic return on investment for the time spent developing the architecture and working to gain a full "buy-in" for the enterprise GIS from the organization.

Enterprise GIS projects reach level 6 (optimization) only with time and patience. GIS planning will accomplish level 6 after the GIS architecture has started to support 14 client cities, a sister company, and the original company. A GIS engineer or enterprise architect position has been established and given a team to continue expanding the system. The enterprise GIS is now fully integrated into daily operations for all 16 supported organizations.

A maturity model created in a spreadsheet program can be converted into a display graphic, by merely highlighting cells in a bright color (light green works well). This transforms each row into a bar of color to display each section's progress within the model. Note that these levels are accomplished over time and not right away. A company might move from level zero to level three in the first six months. However, achieving level four will require a few restarts, a restatement of standards, etc. The maturity model measures the progress of each section through the process of system growth.

7.9 Summary

This chapter covered many of the tools that can crossover from enterprise architecture into enterprise GIS. Readers learned how to establish the goals for a company enterprise GIS. Readers also learned how to define and merge positions between enterprise architecture and a GIS department. They were introduced to the geodatabase as an essential foundation stone for the enterprise-based GIS.

Enterprise GIS goals are an expansion and demonstration for establishing goals will be the first step in providing a strategic vision of an enterprise GIS. Goal statements should be clear and concise on what will be achieved. Once the goals have been determined and listed by priority, the system now

Designing an Enterprise GIS with EA Tools 111

has a direction for growth. New technologies, data, or opportunities can be measured against the goals and determined to be something that should be pursued or avoided.

The chapter covered the concepts of the current state and future state models. The current state is examining and determining what the company has (hardware, software, applications, support, concepts, etc.) at this moment. Future state a vision of where the company would like to become several years down the road. Maturity models monitor and report the company's progress from the current state to the future state.

7.10 Assignment

Assignment 7.1: Build a data, technology, applications catalog

Consider your own company or organization from assignment 3. Using the data, technology, and application catalogs presented in this chapter create a data catalog with a least five sets of data, a technology catalog with at least two computers and printers, and finally an application catalog with at least five software applications. Share and discuss your catalogs with the class.

Assignment 7.2: Catalog analysis

Compare the three catalogs from above and determine if your company has the proper technology and software to support existing or future datasets. (Remember you are planning for data growth over five years.)

Write a one-page paper about your findings and discuss them with your class.

Assignment 7.3: Building a model for current state to future state

Develop a model for the current state to future state for the company for your top five goals. Present the model as if you were presenting at a meeting with company management. Your classmates will fulfill the role of CEO, president, IT supervisor, financial officer. Explain or defend your model to the board.

Assignment 7.4: Build a maturity model

Revisit your company idea from assignment 3. Sit down and create six sections of the company that could work with the enterprise GIS. Select four of these sections and develop a maturity model for the company.

Then answer the following questions:

What is the time from your planning cycle (two years, five years, etc.)?

How long do you expect to reach level 4?

At which level would you reconsider starting a new planning/growth model?

Please present your model and conclusion to the class for discussion.

Bibliography

Baer, D. 2014. "Enterprise Architecture EAOCE." *Enterprise Architecture Center of Excellence*. November 14. https://architecturescoe.org/.

Bedford, D.P.D. 2014a. *Business Architecture*. Kent, OH: Bedford, Denise Ph. D.

Bedford, D.P.D. 2014b. *Business on a Page (BOAP)*. Kent, OH: Kent State University, Dr. Denise Bedford.

Dueker, Kenneth J., J. Allison Butler. 1997. *GIS-T Enterprise Data Model with Suggested Implementation Choices*. Portland, OR: Center for Urban Studies, School of Urban and Public Affairs, Portland State University, October 1.

ESRI. 2007a. *Enterprise GIS for Local Government*. Redlands, CA, December. http://www.esri.com.

ESRI. 2007b. *GIS Best Practices Enterprise GIS*. Edited by ESRI. Redlands, CA: ESRI PRESS, January 1. http://www.esri.com.

ESRI. 2015. *Enterprise GIS. GIS Dictionary*. St. Louis, MO: ESRI, Environmental Systems Research Institute.

Gartner, Inc. 2008. "Gartner Clarifies the Definition of the Term 'Enterprise Architecture'." *Gartner Research* (Gartner), 15. doi:G00156559.

Gill, Asif. 2013. "Defining a Facility Architecture within the Agile Enterprise Architecture Context." *Orbus Software*. Orbus Software. October 1. doi:WP0107.

Graham, Andy. 2012. *The Enterprise Data Model: A Framework for Enterprise Data Architecture*. San Bernardino, CA: Koios Associates Ltd.

Joshi, Prasanna. 2016.. Lecture Data Architecture. *Data Reference Cards*. Kent, OH: Kent State University, Parasanna Joshi, February 18.

Labuschagne, Louw. 2011. "Building Enterprise Architectures for Non-Architects." *Orbus Software*. September 1. doi:WP0011.

Lapkin, Anne, Philip Allega, Brian Burke, Betsy Burton, R. Scott Bittler, Robert A. Handler, Greta James. 2008. "Gartner Clarifies the Definition of the Term 'Enterprise Architect'." *Gartner*, August 12: 1–5.

Michigan Department of Information Technology. 2007. *From Vision to Action: Enterprise Architecture – Strategic Approach*. Michigan Department of Information Technology.

NASCIO. 2015. "Enterprise Architecture Maturity Model." *National Association of State Chief Information Officers (NASCIO)*. January 5. http://www.nascio.org/EA/artMID/572/ArticleID/259/Enterprise-Architecture-Maturity-Model.

Ross, J., P. Weill, D. Robertson. 2006. *Enterprise Architecture as Strategy: Creating A Foundation for Business Execution*. Boston, MA: Harvard Business School Press.

Rouse, Margaret. 2007. "Enterprise Architecture (EA) Definition." *TechTarget*. June 01. Retrieved October 2015. http://searchcio.techtarget.com.

Schedlbaure, M. 2016. "Workflow Modeling With UML Activity diagrams." *BAtimes for Business Analysts*. February 7. Retrieved February 7, 2016. http://www.batimes.com/articles/workflow-modeling-with-uml-activity-diamgrams.html.

Schroede, K., R. Tilley. 2016. "Backcasting." *Design Research Techniques*. February 12. http://designresearchtechniques.com/casestudies/backcasting/.

Schulmeisters, Kar. 2014. "Enterprise Architecture: Bridging Entrepreneurs and Hard Problems." *Orbus Software*. November 1. doi:WP0168.

Sieber, R.E. 2000. "GIS Implementation in the Grassroots." *URSIA Journal* (Vol. 12, No. 1): 15–29.

The Open Group. 19952015. *Enterprise*. January 1. www.opengroup.org/subjectareas/enterprise.

Thomas, Christopher. 2009. "What's Your Definition? Looking at What Enterprise GIS Really Means." *ArcUser* 3032. http://www.esri.com.

United States Geographic Service (USGS). 2016. *Data Standards*. http://ww.usgs.gov/dtamanagement/plan/datastandards.php.

Wilson, North Carolina. (2013). A GIS Manager's dialogue: Is your business semi-integrated or fully system integrated? *North Carolina GIS Conference*, Raleigh, NC.

8
System Visualization

8.1 Strategic architecture

Chapter 8 will cover how various models can be used to create a visual representation of the enterprise GIS. This visual representation or model of the GIS provides enterprise architects, developers, programmers, and users a standard tool for visualizing and planning system connectivity between data, storage, applications, and hardware. These models or diagrams represent the architecture of the enterprise GIS and will become valuable resources. Information is the foundation stone for GIS. Managing information means that we must understand how data connectivity and flow will occur within the system. The information for this model is in our data standards, geodatabase domains, and data catalog.

Up until now, this book has presented information about individual pieces of establishing an enterprise system. This chapter will shift the reader's perspective to a strategic view of enterprise systems. These unique architectural pieces are part of the puzzle that, when assembled, will become the enterprise architecture. This chapter will discuss a strategic overview of the architecture.

All of the information loaded into the models and MS Excel sheets reveals the bare bones of the enterprise GIS. There are meetings to discuss ideas about how to collect, format, use, and store data. There is a realization that the system will have internal or external connections with departments, software applications, and clients. Despite all of the meetings, ideas, and work, no one has an overview or strategic vision of the system. Now is the time to create a visual representation (model or map) of the enterprise GIS.

A visual model of the enterprise GIS provides enterprise architects, developers, programmers, and users a standard tool for planning and displaying system connectivity between data, storage, applications, and hardware (Joshi 2016). Recall the pieces of the puzzle concept about goals, data, and applications. These pieces can have their own "mini" or sub-architecture when these sub-architectures are the pieces of a foundation to build the overview. Models that represent the entire enterprise system (including the sub-architectectures) are known as strategic (or overview) architectures (Joshi 2016).

Strategic architectures are great impact tools for presentations or dramatically explaining to a client why they require the services of a consultant. Reviewing the overview drawings can allow system designers or users to quickly determine potential choke points or blockages of the data flow. A change can be made at this level to open up the choke points and keep the data flowing across the organization.

These drawings of the architectures are important tools for visualizing the path of data through an organization and how it is converted into information. Overall, strategic architecture drawings reveal the enterprise information system in its entirety. Figure 8.1 is an overview of a proposed enterprise GIS for the Ohio Turnpicke Commission (OTC). There is a large amount of detail shown and communicated to the viewer. These system drawings can provide a large amount of information in an easy-to-understand format.

Overview drawings can be used to educate people about the inner workings of the enterprise information system. The lines, arrows, and constructions display the paths, travel directions, origins, and final destinations of data within the system. Sub- and strategic architectures are maps of the system process that transform data into information. These system maps enable people to see, follow, and understand how their information is created, stored, and eventually accessed by the users.

Strategic architectures can be used as financial planning aids for one or more fiscal years. Sub-architecture (plan) views are the key to using a strategic architecture for fiscal budgeting. Every sub-plan represents a piece of the overall system. Sub-plans can represent when a future capability or asset will be acquired by the system. These future sub-architectures must include the following information: year of expansion, estimated costs, demonstrate connections and benefits to the existing system. A fiscal officer will support an enterprise project's construction, maintenance, and expansion if they understand how it will impact the budget. These sub-architectures are visual representations of smaller subsystems within the overall enterprise system.

Sub-architectures should be considered the bricks for building the overview architecture. Each sub-architecture is a detailed visualization of a small part of the enterprise system or architecture. An enterprise system should be developed brick by brick. A geodatabase sub-architecture would reveal what feature datasets (folders) it will contain and what feature classes (files) will be found in the feature datasets and are found inside the geodatabase. When developing the enterprise system, each small piece should be developed as a sub-architecture, and then connected to the other "bricks" to create a strategic vision of the entire enterprise system. These sub-architectures are connected together, to create the strategic architecture or enterprise vision.

The current state to future state models is an excellent example of building a strategic architecture from the subsystems. Using the inventorying of the currently available data, applications, hardware, and processes, a designer

System Visualization

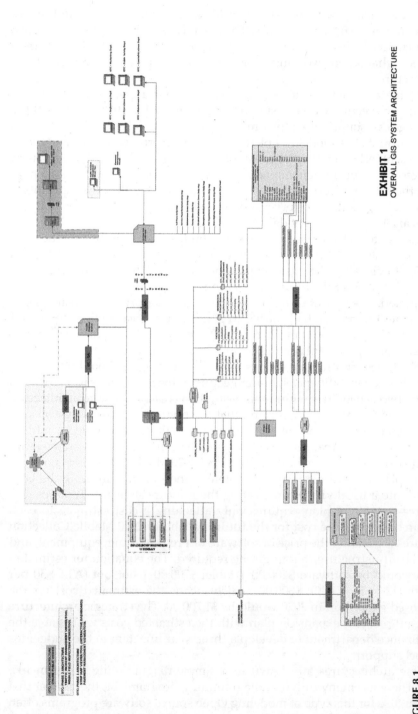

FIGURE 8.1
OTC overview architecture. (Author's project files.)

makes the current state model. When the model is finished, it can transfer into an ongoing state architecture. The current state architecture is simply a visualization of the current state model. People always find a few surprises between what is perceived and what is the actual situation in a review of this architecture.

Drawing a current to future state architecture is a great example of how the sub-architecture works to support a strategic architecture and how these drawings can support fiscal planning.

The current state model is a picture of the enterprise system as it exists in the present: a listing of what capabilities and assets that the system has in the present. A future state model is about what people expect the system to be in the future. A future state vision will show what the new capabilities, skills, and assets the enterprise system will require in a year, two years, or five years, from the present day.

Future state architectures are the key to budgeting funds to cover the costs of future system builds. The aforementioned sub-architectures are what will be used to visualize the future expansions of the established enterprise system. Do not restrict the future state visions to only one year, think in a five-year period. The City of Oberlin established an enterprise system and, eight weeks into the project, had started to plan for future capabilities for Year 1 after build: stormwater control; Year 2 after build: water department service valves; Year 3: pavement condition indexes and construction history.

Each of these future state visions for Years 1 through 3 should refer to one and only one capability per sub-architecture. If the future state is referring to adding two capabilities within that year, then create a separate architecture for each new capability. Assuming that the original enterprise system was implemented in 2019, each of the future sub-architectures should be entitled using this formula Year + space + capability. The first future sub-architecture planned after the system build would read 2020 Mobile Data Collection.

This plan for a three-year expansion of the system can easily be converted into a fiscal vision. Converting these future state architectures into a three-year fiscal plan requires only attaching a cost estimate to each sub-architecture. The cost for the future capability 2020 Mobile Collection requires estimating the price of software or programming, equipment, and new IT infrastructure that might be required. The equation for estimated costs would be software $400.00 + tablet $700.00 + hotspot (ATT $50 per month ($12 \times 50 = 600$)) $600.00, which means the estimated cost for the proposed expansion in 2020 would be $1,700.00. The strategic architecture with a three-year expansion plan with the estimated costs will enable the city finance department to develop a three-year line item in the budget for project support.

These architectures are drawn in a simple diagramming program like DIA. There are many different diagramming programs on the market that one can use for this type of modeling Open source software programs often have many of the same basic features that are provided in professional or

System Visualization

licensed software applications. When one is doing a simple diagram, all the bells and whistles of the "professional" version are not necessary. However, it is crucial to find a program that you are comfortable using because these architectural drawings can get complicated.

Figure 8.1 shows a complete enterprise GIS architecture. Future items need highlighting, and the future architectures have a corresponding highlighted in the same color. This technique allows the viewer to determine the present from what is the future quickly. A future state is a vision of what the enterprise GIS is expected to be several years in the future. Future state architectures are excellent tools to represent system planned system growth graphically (Labuschagne 2011).

These models or diagrams represent the architecture of the enterprise GIS and will become valuable resources. Information is the foundation stone for GIS. Managing information means that we must understand how data connectivity and flow will occur within the system. The information for building this model is in data standards, geodatabase domains, and data catalog. Refer to Figure 8.2.

This model reveals how the geodatabase's feature datasets (main folders) will categorize coverages, facilities, information, and OTC data. Each category is a grouping of related data classes divided according to what information is inside each class. Note that the feature class Fac_Toll_Plaza is in the feature dataset facilities. It is placed in this dataset because it contains data on Toll Plaza facilities. This model also notes that a geodatabase can hold any number of feature datasets, and these cannot exceed 1TB (ESRI 2007).

The data connections between applications, hardware, and the geodatabase are diagrammed in a model of connectivity. Figure 8.3 is an example of a data connection model. This model was built to show how pavement data travels from legacy datasets, ArcGIS (desktop programs), pass through a quality control check, and then be loaded into the feature class. Note how the data format standards are used to identify how each piece of data will fit into the feature class. System models are diagrams that provide a visual overview of the entire enterprise GIS. The field of enterprise architecture calls these diagrams or models architectures. There are two types of architectures: overview architecture and sub-architectures.

8.2 Sub-architecture

The sub-architectures enable users to "zoom in" to a specific piece of strategic architecture. Recall the parts of the puzzle concept about goals, data, and applications; these pieces can have their own "mini" or sub-architecture when these sub-architectures are the pieces of a foundation to build the overview. Models that represent the entire enterprise system

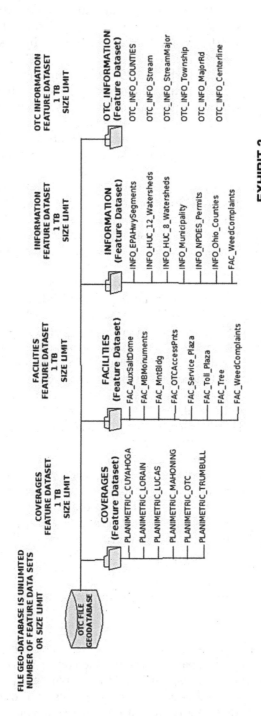

FIGURE 8.2
Geodatabase architecture model. (Author's project files.)

System Visualization

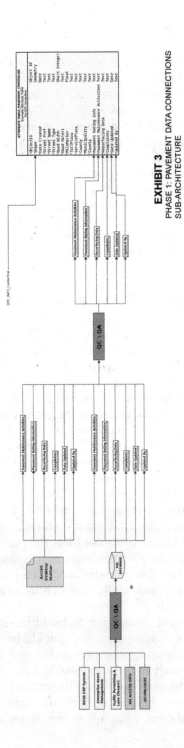

FIGURE 8.3
Data connections model. (Author's project files.)

(including the sub-architectures) are known as strategic (or overview) architectures.

Sub-architectures (subsystems) are used to visualize individual subsystems of the architecture. These smaller subsystems can then be joined together to create a strategic view of the entire architecture. Consider the sub-architecture show in Figure 8.3 entitled Phase 1: File geodatabase sub-architecture. A geodatabase sub-architecture displays the internal storage of the file geodatabase. The geodatabase is a cylinder left to the file folders representing different classes of information. This subsystem visualization shows that the geodatabase contains four data classifications, and each one is limited to 1 TB in size. These lists within the classification provide examples of what the data is, and how to name the dataset. The second folder is entitled "Facilities" and is considered a feature dataset. Individual data files use this naming convention, Class_Data. This is demonstrated by looking at the folders Facilities, where the first dataset is FAC (facilities)_AuxSaltDome (Auxiliary Salt Dome).

Find the very last classification type in the above geodatabase (OTC_INFO_Centerline), and then look at Figure 8.3. Note the line and arrow that points to the word shape table ATTRIBUTE TABLE_PAVEMENT_CENTERLINE. This shows that the table supports the centerline dataset. The lines with arrows show the route of travel for data flow in this subsystem. There are two origins for the data about the centerline. A desktop station is feeding the following **Insert** information Pavement Maintenance Activities, Pavement Rating Information, Resurfacing Data, Complaints, Date updated, and finally Updated By into a QC\QA process. After QC\QA, the data is shown connecting to and replacing data inside the attribute table to the corresponding data connections.

The second route starts from a set of existing systems: Ross ERP, Enterprise Asset Management, Traffic Permitting, and Lane Closures, MS Access, and Downloads data from these systems is passed into a QC\QA process that removes errors, updates formatting, and ensures accuracy. These systems are passing Pavement Maintenance Activities, Pavement Rating Information, Resurfacing Data, Complaints, Date updated, and finally Updated By into a SQL database. This intermediate database is a storage unit that passes the information into the same QC\QA process from the paragraph above. This second QC\QA process ensures this information is compatible with the new system, merging the two data streams, and then passing the information over to the centerline attribute table. Sub-architectures are the building blocks of strategic architecture.

Sub-architectures will allow a designer to build the strategic view, one small step at a time. The strategic view can quickly become very complicated and confusing. System designers must strive to avoid getting lost in the details. Using the sub-architectures to visualize the section details and then use the detail pieces to complete the architectural puzzle, will prevent designers from getting lost in the details. Build the strategic view one piece at a time, work at a tactical level, while using the strategic level to keep the system on the proper course.

8.3 Summary

Chapter 8 discusses how to visualize enterprise systems through diagrams (maps). "A picture is worth a thousand words" is quite applicable in today's world of complex information systems. A chart, diagram, picture, or any graphic will aid system designers, implementers, users, and management while communicating to avoid getting lost in the details. A person can quickly pinpoint where the system connects with users, data, clients, or external systems. The company's path from the current state to achieving its future state or ideal vision is displayed to all stakeholders.

Enterprise or system architecture can (and often will) have a high level of complexity. When building from a current state, it is wise to use sub-architectures to visualize the details of each planned expansion. These small sub-architecture building blocks are the foundation of the overview. This overview or strategic level architecture provides a high-level view of the entire enterprise system. Strategic and sub-architectures are the most effective tools for visualizing the development and growth of the enterprise system.

8.4 Assignment

Assignment 8.1: Visualizing the entire system

The following exercises can be completed with graphics software. However, it is recommended that a diagramming program is used to finish the assignments.

Revisit your company or organization's current state to future state model and the maturity model from assignment 7.

Assignment 8.2: Visualization of current state

Examine your organization's current state to future state model. Now develop a strategic architecture visualization that represents the current state of the company.

Assignment 8.3: Visualization of sub-architecture

Examine your organization's current state to future state model. Select one item that will be added or developed for the system at the midpoint of your development cycle toward achieving the future state.

1. Create a sub-architecture that reflects how this piece will expand the current state architecture.
2. Show how and where it connects to the current state architecture.

Assignment 8.4: Visualization of future state

Examine your organization's current state to future state model. Now develop a strategic architecture visualization that represents the future state of the company. If you are following a five-year planning cycle, then this architecture should reflect that it is five years from the current state. Hint: Develop one or two more sub-architectures in addition to the one from Figure 8.2.

1. Show all of the new sub-architectures and how they connect to the existing current state.
2. Show how each of the sub-architectures connects within the future architecture. (Note it is possible that they might not interact with each other.)

Please present results to the class for presentation and discussion.

Bibliography

ESRI. 2007. *Enterprise GIS for Local Government*. Redlands, CA, December. http://www.esri.com.

Joshi, Prasanna. 2016. Lecture Data Architecture. *Data Reference Cards*. Kent, OH: Kent State University, Parasanna Joshi, February 18.

Labuschagne, Louw. 2011. "Building Enterprise Architectures for Non-Architects." *Orbus Software*. September 1. doi:WP0011.

9
Governance

9.1 What is governance?

The new enterprise system has been implemented and accepted. Now the organization must safeguard the system from corruption. Protecting against data corruption is done through system governance! Governance is an essential part of any enterprise system, for this is how system principles, goals, standards, etc. are enforced. Without governance overseeing the system and monitoring growth, the system would quickly descend into chaos (Bedford 2014a).

One can look at system governance as a type of management. Governance is responsible for ensuring that architectural policies are understandable and carefully followed. Management will ensure that the organization progress from the current state to its future vision (state) without detours, dead-end development, or being ignored. Everyone in the organization shares the responsibility for system governance.

There are some basic requirements for system governance to be effective. Governance must have excellent leadership; a person who can communicate with everyone sets an example with the classic concept, "I would not ask you to do something, that I would not do" (Bedford 2014a). Influential leaders know how to avoid confrontation by calmly listening, thinking before acting, and avoiding the quick jump to a conclusion. They reinforce the concept of a team working together, yet have the wisdom to assign a solo project to calm a situation. A good leader enables his team to succeed, communicates through listening, understands people, stands firm, yet knows that sometimes it is better to compromise.

Effective management or governance is the result of proper organization. The organization does fall under the jurisdiction of leadership. Well-organized governance provides guidance, ensures adherence to the architectural principles, and above all promotes one message. A disorganized team could create conflicting messages about how to proceed, or what is acceptable will setback any effort at keeping the system on the path to achieving the vision outlined in the future state model. The system avoids conflict and is responsible for what, which area of the system, and how to respond if something violates system principles.

The leader of the governance team must also comprehend how the company operates, which department (business unit) is responsible for a particular section of operations (Ross et al. 2006). It is essential to know the head of a department, their assistant, and the supervisor. Understanding the internal workings of these business units will aid in knowing how to apply the governance policies that are specific to each group.

Governance is sometimes applied to a sub-architecture that stands alone, like data architecture. A data architect will establish governance to manage the collection, storage, formatting, access, and use of company information. Data governance is the first line of defense against data corruption or loss. Data governance is also a subset of enterprise or system governance.

Case: The City of Oberlin, Lorain County, Ohio

The primary reason the city's IT manager initiated the enterprise project was to organize and govern the city's information and GIS efforts. The Public Works Department is the main warehouse of information in the city and had started a GIS program. The city-owned electrical utility Oberlin Municipal Light and Power Service (OMLPS) was using a limited GIS for utility pole information. Several other departments were also interested in developing GIS capabilities.

City departments depended upon the Public Works/Engineers departments to provide information for construction, road improvements, and sanitary and storm sewers for ongoing projects.

The information in Public Works had not been organized into a coherent library. A lot of the city's infrastructure information was contained in a 2006 MS Access database and this database was on a very old machine. There were several AutoCAD drawings that held most of the city's line (sewer, water, and road), point (catch basins, addresses, sign inventory), and polygons (buildings, parcels, catch basins). Trying to access or update this information represented a challenge.

The author held several discussions with the IT manager and city engineer. The city decided to move forward with the idea of developing an enterprise GIS over an enterprise architecture foundation. The first challenge was with wrestling the data into one set of formats and procedures. After an initial kick-off and first enterprise planning meeting, the decision was made that all the city's data would be placed into one online GIS program housed in Public Works and shared with all city departments.

A fully developed data architecture was soon discussed, approved, and put into place. A data management and sharing plan was established under the umbrella of data architecture. This plan specified that all city information would be housed in Davey Resource Group, Inc. "DRG" GIS program Asset Manager.

Enterprise governance is the management of a system from a strategic viewpoint. A significant goal of governance is developing clear communication between all levels of the organization and system. Any governance policy must be clear, concise, and easily understandable by everyone in the organization (Graham 2012). Anyone, when asked, should be able to explain their role and responsibility under the system's governance policy.

Successful governance has a foundation of principles. These principles are responsibility, accountability, transparency, and impartiality (Bedford 2014). When someone agrees to fulfill a position in the architecture, they are now responsible for "getting the job done right!" which stands for the person taking no short cuts, working hard, and sometimes making hard decisions. Accountability follows responsibility because it means that people are to be accountable to the system, company leaders, and fellow employees. The concept of accountability and responsibility together is demonstrated by this sentence "if you make a mistake, admit to (own) the mistake, don't blame others, learn from the mistake, and continue." When people realize that there will be accountability system errors decline, and they pay attention to doing the job right.

Accountability with responsibility creates a foundation for transparency and impartiality within the architecture. Transparency means that everyone should be aware of what is happening, who is responsible, and why and how it is happening. There should be no secrets, and no one should feel left out of any conversation. Transparency will only occur with clear, concise, and immediate communication between departments and staff. Any issues that are left to rot or boil can and will harm the idea of transparency. Once transparency is lost, then trust between employees and the system will also be lost.

Case: The City of Oberlin, Lorain County, Ohio

A rather unique governance structure has been put into place over the Asset Manager program. Public Works IT manager has been established as the overall Asset Manager administrator, each city department director will become a sub-administrator, responsible for his or her department's information set. Each administrator will have the right to add users, determine user privileges, and decide what information will be shared. This arrangement allows departments to share, review information, and prevent duplication from occurring in the city's information system.

Sub-administrators can only grant editing permissions to their set of users. This means that each department can work independently to gather and edit information without the worry of deleting or overwriting another department's file. Here is an example of the system in action. OMLPS must replace three utility poles along Main street. A crew is sent out and they quickly access Public Works infrastructure information from the field to determine what sewers or watermains might interfere with digging out the poles. Crewmembers open the OMLPS layer for utilities, mark the poles replaced, and save the information. While they are on the job, they notice that a nearby hydrant is leaking due to a loose cap. They tighten the cap, alert the Water Department to the situation, and provide the hydrant's ID and location from their tablet and then move on to their next job.

A member of the Water Department quickly queries GIS by the hydrant number. He or she reviews the maintenance records, which reveal that the cap is scheduled for replacement. They load up a truck with the proper tools, new cap, drive out to the hydrant, make the repairs, and update the hydrant's maintenance record from the field. The online GIS aids departments to share information, perform edits, and keeps all updates in real time.

Designing, building, along with implementing the enterprise architecture is one thing: having people accept and support the system is another (Ross et al. 2006). The ultimate success of the system depends upon achieving "buy-in" or belief of the staff that they own or have a voice in the system. Impartiality is the key to achieving "buy-in" from the team. Everyone should have the same standards of accountability or responsibility, and there are no favorites or "just letting things slide" not even once. When ideas surface, each suggestion should be judged equally by merit. All designs are discussed, nothing should be dismissed for reasons like "it is stupid" or "that person is unpopular." Everyone should have a voice, listen, and receive a fair hearing. Clear communication, transparency, responsibility, accountability, and being impartial, will obtain the trust or "buy-in" from the staff.

Case: Medina County Health Department (MCHD), County of Medina, Ohio

The author experienced a nice solution to communication problems at his time with the MCHD. Each department would hold weekly meetings where staff could simply talk and learn. These weekly meetings were part of the office routine that made everyone comfortable. Staff members would hold conversations on various topics and simply relax. The meeting was considered informal, so people could address issues that bothered them without having to involve HR. This allowed the staff to clear up misunderstandings or settle minor issues before they became a serious problem.

A weekly meeting also serves the purpose of ensuring that everyone feels that they belong to the group and people will listen to them. The author would attend these meetings to answer questions about the enterprise GIS, data concerns, or just to listen and learn what they were thinking. Attending the occasional meeting reinforced the message that the author was easily accessible and friendly enough to engage in a friendly chat. It is truly amazing how much information a friendly listener can glean about the staff's perception of a project in an informal setting. The author highly recommends to any reader or consultant not to pass up on this type of opportunity.

9.2 Governance policy

The governance or architecture policy is not a simple document, and it is made up of all of the processes of system governance. All of the principles, goals, standards, etc. that were discussed, examined, defined, selected, and agreed to, by the architects, system designers, management, and staff are parts of the policy. The architecture policy is a group of policies working together to ensure that company activities comply with enterprise architecture.

These policies, when supported by principles, will supervise a three-way relationship between the architecture of the system, organization, and external connections. Managing or balancing a principle or policy when a conflict arises is one of the functions of governance (Labuschagne 2011). There will always be a compromise made in larger systems between policy and stakeholder. Managing the architecture processes will always be a balancing act between what stakeholders perceive, a policy, and the principle.

A successful architecture will grow, encourage ideas, and continue progress toward the future state. Principles, goals, processes, etc. might have to be modified to accommodate any new expansion of the system. Successful growth will encourage new ideas about how the system can be improved, expanded, or even applied to new opportunities. The new system and architecture have been accepted and are now increasing, maybe, too fast. An active governance policy and review will ensure that this rapid expansion does not derail progress toward achieving the desired future state.

The business architect (BA) or whoever is fulfilling the position (GIS manager, assistant manager, coordinator, analyst) is mainly responsible for system governance (Bedford 2014a). The BA will create a series of checks and balances to oversee the creation and monitoring of the architecture system. The BA must develop workflows, methods, inspections, and a review process to ensure that the architectural policies followed. They are also responsible for ensuring that all of the components of the system are accountable to company oversight (Bedford 2014a).

The BA is not expected to create the entire governance policy and review it from a vacuum. The first tool is the enterprise architecture design itself! The design is an essential resource because it displays and records the progress from the current state to the future state (ultimate goal). Other tools are the records or documentation about discussions, reviews, system updates, and reviews that have impacted or will impact the system. A successful enterprise is dynamic, changing, and adapting, in many ways, it is a "living" system. The challenge is to develop a network of governance process that encourages and not hamper system growth while keeping the architecture on the path toward achieving the future state.

It has been the author's experience that many levels of local government overlook codifying their policies into one governance document or practice. A few people will refer a consultant to a federal document, and they might even have a copy of the document at their station. This is a normal situation, because most organizations feel that they don't have the time or resources to assign a person to manage the governance of the enterprise system. There is also the perception that it is a departmental issue and falls under the responsibility of a supervisor or director.

A consultant must work hard to help establish good governance policies and enforcement habits with their clients. The lack of a strong governance effort will result in the whole enterprise system drifting onto a divergent path. These divergent paths will bypass all of the principles, standards, and procedures and replace them with corrupt data, duplicated pathways, along with redundant projects. Basically, within two or three years the gains made by the enterprise system will be lost, and the organization will repeat the whole process all over again.

Case: City of Strongsville, County of Medina, Ohio

The City of Strongsville's engineering department consisted of three people: the engineer, assistant engineer, and a GIS technician. The engineer quickly grasped the concepts of system governance. He had championed the GIS project to other department heads and quickly took ownership of the idea of system governance. He designated his assistant engineer to oversee developing governance policies with other departments. The assistant engineer and GIS technician began hosting meetings with other department staff to work out a governance policy for the city.

There was no real push back from the other departments about forming a governance committee and policy. City staff moved quickly to establish standards, principles, and methods as part of their new governance efforts. One of the first items that changed for the public was the submission of developer's plans to the city for review. The city switched the requirements from filing only filing paper plans to asking for plan sheets in a digital format of GIS, AutoCAD, JPEG, or PDF. Requiring digital versions of the plan sheets allowed the staff members to quickly update the GIS datasets, after the developer's plans were reviewed, approved, and accepted by the city.

9.3 Governance review board

A governance review board not only ensures compliance within the existing system, but also ensures compliance with the system architecture by new ideas or processes (Bedford 2014). A review board will review proposed concepts, processes, methodologies, etc. to ensure that it adheres or further develops system capabilities within the existing architecture. Reviews will be performed quarterly to ensure that there is compliance with policies, standards, and overall architecture. Quarterly reports will form the foundation for the acceptance, adjustment, or rejection of an item by the architecture review board.

The BA and architecture review board will work together to develop a methodology to review items or ideas that comply with the architectural policy. The current state to future state model, system documentation, and the architecture (system) principles, etc. will form the foundation for this compliance review. Changes, ideas, or solutions to problems, must be reviewed to determine their impact on the enterprise system.

A review board will ensure that the company's actions, decisions, and development all comply with the architecture policy. The board should have one representative from the executive, data, technology, process architect, supervisors, and finally the business architect. The committee will conduct policy reviews every quarter and approve or deny compliance waiver requests. The BA's task is facilitating board meetings and reporting decisions to the enterprise architect or company stakeholders.

The governance or architecture review board should always decide through voting, with the majority rule winning the vote. A review board should not perceive themselves or be heavy-handed enforcers of the architecture's policies. The board should have an odd number of representatives. The goal of the board is to help everyone think in terms of the enterprise system when making decisions (Bedford 2014a).

Many people in local government view a review board as just another committee as a waste of their time. Again, it is the consultant's responsibility to establish good governance habits, and to ensure that a review board is formed. There will be resistance to establishing a properly functioning board. The root of this resistance is not found in staff resentment of the system, it comes from the feeling that a board is not important and will take up too much of their time. A key to overcoming the resistance is to locate and place staff members who are experienced with technology, information systems, and are excited by the idea of improving their enterprise system.

Case: The City of Oberlin, Lorain County, Ohio

A little earlier in this chapter we discussed the City of Oberlin's arrangement of multiple administrators for their online GIS. The administrator setup will be: Public Works IT manager is the overall Asset Manager administrator and each city department director (or a supervisor) will become a sub-administrator who is responsible for the department's information.

The city staff decides that these administrators will also be included on the review board. The position of review board chair went to the Public Works IT manager. The other review board members are Water Department supervisor, city engineer, Storm Water coordinator, and a representative from the city administration. This board was formed and immediately started a cycle of monthly meetings to review ideas, new technology, and methods suggested by employees or citizens.

Oberlin Municipal Light and Power Service (OMLPS) had the first appearance before the review board. OMLPS wanted to use an iPad equipped with a Bluetooth Trimble GPS for field data collection, and then build a sub-system to upload the data into the new Asset Manager program. Members of the review board had some good questions about the new idea.

First, they wanted assurances that this hardware would be used only once and then placed on a shelf. Second, OMLPS had to assure the board that the software package could indeed gather field data and upload it directly into the enterprise GIS. Board members asked if the Bluetooth device provided the same accuracy level as a normal handheld unit. OMLPS staff members informed the board that there was a loss of several inches in positional accuracy, but the points were still accurate in the sub-meter range.

The review board considered that the data to be gathered would be used for planning and informational purposes. Therefore, it did not require a certified surveyor's accuracy to be plus or minus one inch. Board members also agreed that the ability to edit and upload data from the field into the enterprise GIS was a nice feature.

There was a cost-saving element involved in the project. Editing data in the field and then immediately uploading it into the enterprise GIS saves staff members time. A 10-minute editing and uploading session took the place of an hour, maybe two because the crew did not have to travel back to the office to perform the work. Time saved equals money saved, because the city no longer paid for time lost due to travel and performing data edits in the office.

9.4 Compliance waivers

When a proposed idea, methodology, process, etc. might not fall into the enterprise system's scope, the review board can issue a compliance waiver. The architecture review board can grant compliance waivers based upon the recommendation of the appropriate board member and if it can be proved to be in the best interest of the company or the enterprise system (Bedford 2014). The review board can place the time of expiration on any waiver that is accepted. Establishing a time limit will ensure that the item returns to the process for reevaluation for compliance.

Listing the requirements for compliance will be an essential tool used by the board when considering a case of conformity (Minoli 2008). This list should include a description of why the item is out of compliance, what will be required (milestones and deadlines) to comply, the actions to be taken to correct the problem that would achieve compliance, an explanation of the consequences if compliance has not occurred, and, finally, what happens if the agreed-upon deadline for compliance is missed.

The following is an example of the compliance requirement list.

List of requirements for compliance

Extent

The proposed methodology does not comply with the established data collection process.

Items required for compliance

1. A written explanation of why this format should supersede current formats or why the board would accept a new format.
2. A written report is due in two weeks from today.
3. A successful test of the new data format deployed in a small pilot project.
4. A report is due in three weeks after item 1 is complete.

Actions needed

1. The board will review all the results from the pilot project.
2. These results reviewed by all departments.
3. If accepted, then the board will vote to add the new format to the list.

What are the consequences?

If the format is used by only one particular project, then that will impede the sharing of information between departments, maps, and staff. A proprietary format can lead to a loss or corruption of data and delays in completing assigned tasks. These delays could result in failing to meet client deadlines, potentially creating the loss of projects and revenue.

Penalties:

If the format is not brought into compliance by the agreed-upon date, then it will be rejected, and all projects using the form must change to one in accordance or face termination.

Sample response to the compliance waiver

Extent: Acquiring field data in an emergency using a drone.
Items required for compliance:

1. During the emergency, a drone can fly into dangerous areas and perform a search, data acquisition without endangering emergency personnel. We can use a live feed from the drone to help emergency personnel determine if there is a person who is trapped and must be rescued. Four types of data will be downloaded from the drone.

Video, Raster, Vector, and GPS (X, Y positional) video can be captured and then reviewed on site or at the office.
2. We will provide a written report after two weeks of testing the drone's abilities in small areas and mock emergency drills.
3. We will provide a presentation and demonstration of the drone's abilities to the board for their approval.
4. We agree to provide a written report within one week after approval.

Actions needed

1. The board will review all the results from the pilot project.
2. These results will be reviewed by all departments.
3. If accepted, then the board will vote to add the new format to the list.

What are the consequences?

If the drone fails during an emergency then it could lead to a loss of data and potential harm or death to emergency responders. Damage to a camera or sensor could leave the drone functional but unable to respond to the pilot, which would result in the loss of an expensive piece of equipment.

Penalties

If the drone fails to operate properly in the two-week testing period it will be withdrawn from consideration as a tool during emergency response.

Compliance waivers are a very useful tool to establish if there is a need to consider changing the governance policies to accept the new tool or data type. These waivers are in many ways a pilot project. Like a pilot project a waiver has a limited work scope and a strict timeline for completion. People in local government are reassured by those limitations, everyone worries about a small project suddenly spinning up to a monster that simply sucks money and resources from the organization. Compliance waivers are very useful for identifying new methods, equipment, and ideas that might benefit an organization or the enterprise system.

Case: The City of Avon, Lorain County, Ohio

The service department of Avon was unsure of the method proposed by CVE and the author for digitizing fire hydrant locations from an aerial photograph. Although there was not a written set of compliance waiver requirements the service director laid down a few verbal requirements. He limited the project scope to four city streets of his choice. Second, the accuracy level had to be within one foot plus or minus six inches. This would be tested by him and the GIS specialist in the field with a tape measure.

Digitizing hydrant locations from the four streets was accomplished with a couple of days to spare. The accuracy test was next, and the data passed with flying colors. CVE's GIS team achieved an average accuracy level of just under one foot. The service director was satisfied and a full-scale project began the next week.

This compliance waiver and test was not wasted time. The service department and CVE's GIS team learned a couple of lessons. Lesson number one is that shadows can have a negative impact on digitizing efforts. This could be overcome by using a different aerial when available or changing the color saturation values. The second lesson is that hydrant locations might not be accurate in the information files provided by the previous contractor. High definition aerial photographs and plan sheets were to be considered the primary sources for location information for the city hydrants.

Any information system requires proper management to be successful over time. The field of enterprise architecture calls this management governance. It is responsible to make sure that the established principles, policies, standards, and methodologies are properly followed. Proper governance will ensure that the enterprise project does not become sidetracked or travel down a divergent path. The responsibility of governance does not fall on one or a small group of people; everyone is responsible for system governance.

A leader of the governance should be able to communicate with everyone. A good communication can avoid confrontation by calmly listening, thinking before acting, and avoiding the quick jump to a conclusion. They remain calm and use management techniques like assigning a solo project to separate people and allow a hot situation to cool. A good team leader communicates through listening, accepts opinions or ideas, can be flexible or stands firm when required, and understands the value of compromise. A true leader enables and pushes his team to success.

This type of effective leadership is effective for managing and creating a successful governance effort. Good communication and organization skills are a must for establishing and managing an enterprise effort. Well-organized governance provides guidance, ensures adherence to the architectural principles, and above all promotes one message. A disorganized team could create conflicting messages about how to proceed or what is acceptable will set back any effort at protecting the enterprise system from mistakes or dead-end developmental path. Effective governance is the result of proper organization and communication.

9.5 Summary

This chapter answers the question "What is governance?" The chapter then explains that the role of governance is to ensure that progress continues toward achieving the vision of the future state. Governance ensures that the system does not drift onto a dead end line of development. It acts as a navigation aid, ensuring that all efforts are concentrated and working together to achieve the vision of the future state.

The business architect or whoever is fulfilling the position (GIS manager, assistant manager, coordinator, analyst) is mainly responsible for system governance. This person will convene a review board that conducts regular reviews of system developments and ensure that everything complies with established architecture policy. A successful business architect will develop a system of governance process that encourages and not hamper system growth while keeping the system on the path toward achieving the future state.

9.6 Assignment

Assignment 9: Define governance

Write a one-page essay that answers the following questions:

1. What is governance?
2. Why is governance important?
3. How would you establish a governance policy?

Please present results to the class for presentation and discussion.

Bibliography

Bedford, D.P.D. 2014. *Business Architecture*. Kent, OH: Bedford, Denise Ph. D.

Graham, Andy. 2012. *The Enterprise Data Model: A Framework for Enterprise Data Architecture*. San Bernardino, CA: Koios Associates Ltd.

Labuschagne, Louw. 2011. "Building Enterprise Architectures for Non-Architects." *Orbus Software*. September 1. doi:WP0011.

Minoli, Danile. 2008. *Enterprise Architecture A to Z: Frameworks, Business Process Modeling, SOA, and Infrastructure Technology*. Boca Raton, FL: CRC Press, Auerback Publications, Taylor & Francis Group.

Ross, J., P. Weill, D. Robertson. 2006. *Enterprise Architecture as Strategy: Creating a Foundation for Business Execution*. Boston, MA: Harvard Business School Press.

10

Conclusion and Future of GIS

10.1 Conclusion

One curious person seeking to find a better way to handle project mapping and data analysis can trigger a rush toward an enterprise GIS. This curious person succeeds in supporting a project with a beautiful map that impresses co-workers and management. Soon this person is now doing mapping for other departments and helping other staff members learn how to work with GIS software. There is no plan, no coherent system, just a small group working with GIS. Yet, this small group of people who are sharing files, ideas, and making maps have formed an ad hoc enterprise GIS.

There are those who feel that an enterprise GIS is simply providing more than one user, one department access to GIS tools and methods. Each person or department has its own GIS software package and dataset on a server. Every desktop machine has a copy of the main geodatabase along with aerial photographs, small department projects, and a couple of multi-department maps too. Each user's data and maps are backed up to individual portable hard drives.

This new "enterprise" GIS has many issues. Issue number one is the large amount of duplicate information that is simply taking up valuable space in memory and storage. There are redundant maps and efforts where one or more of the departments have produced the same map or dataset for similar projects. Due to a lack of IT support the system response is slow because of choke points in the data flow, outdated infrastructure, and poorly constructed pathways to files on hard drives. No person or department is enjoying a benefit from this system.

When there is no benefit to the entire organization there is no enterprise system. Enterprise GIS can become a true enterprise system through the incorporation of ideas, tools, and methods from field of enterprise architecture (EA). When a GIS follows an ad hoc growth and development style, the result is a system in chaos. It is a system filled with errors, duplicate data, efforts, projects, and viewpoints that lacks any cohesive unity or cooperation. All this chaos results in lost effort, patience, data, and time. Remember the old saying "time is money," lost time simply means lost money. Money or profits lost in this fashion cannot be recovered.

Businesses, non-profits, and local governments have all come to realize this fact about money losses due to lost time. There has been a push to find better methods for collecting, creating, storing, using, and managing data or information. The big buzz word is "enterprise system"; this was connected to GIS for yet another buzz word "enterprise GIS." It seems like people will always chase a buzz word especially when it is perceived as a "hot" item or an easy fix to a problem.

There are many software packages promising quick or easy solutions right out of the box. When a software application is purchased, the purchaser suddenly discovers that the software doesn't quite fix their problem. A quick phone call to the tech support and the response is that the software isn't really designed that way and for an additional fee, a customized solution is available. The author has had to step into situations where less than reputable consultants have promised easy fixes, then taken the money and fixed nothing.

The truth of life in the world of information management is that there is no easy quick fix available for system design. Successful, efficient, and stable enterprise systems are the result of a lot of time, effort, and dedication by the project team. There are no short cuts or fast lanes when developing these enterprise information systems. Establishing and maintaining a successful enterprise GIS or EA requires dedication and a major commitment of time, effort, and patience.

The ideas, concepts, and methodologies from EA are an effective way to create a unified, organized, and efficient enterprise GIS. Using concepts, methods, and ideals from EA for enterprise GIS should be done by everyone designing or developing plans for a GIS. Merging EA with enterprise GIS requires more thought and effort, but this extra level of planning will simplify system implementation and maintenance. The return on investment (ROI) of extra effort is a well-organized, efficient, responsive system that converts data into valuable information.

People ask the author a very simple question, "Where do we begin?" and the answer is, determine why or what precipitated the desire for a new information system. Clients have responded with a variety of reasons like we lose data, or the data is bad. We have lost information about the city's infrastructure when people retire. The system takes forever to respond to a simple command or data request. The list of reasons seems endless, but it all comes down to lost data, lost time, and lost effort, because the system is disorganized.

There is no set starting point when designing an enterprise system. One good starting point is discovering the business capabilities for an organization. This is part of system planning; however, it is essential to remember that capabilities could change as the new system goes online. A business capability model reveals WHAT the business creates, or performs that brings value to the company, stockholders, and personnel. The business capability describes the stable elements of the business and provides an important layer that aligns business and resources. It is hard to design a system that supports capabilities if you do not know what are the business's capabilities.

The author prefers to start the design process by establishing goals for the organization. Goals are a vital piece of the foundation of the enterprise GIS. Clear, concise, well-defined goals enable people to envision where a company could be in the future. This vision of a company's future is known as the future state. The goals that are to be completed in the future will enable the company to achieve this vision of the future.

An assessment of a list of goals will determine which ones are primary or secondary in importance. During the goal assessment when everyone agrees on a goal, then it should be considered a primary objective. When a potential goal tends to support another goal, then you can safely assume that is a secondary goal. Secondary goals act as supporting steps that can guide a company to completing the primary goals. A common mistake is to overlook this interaction between goals.

Completing several step goals that relate to a final goal can be tracked in a system progress model. This enables designers to measure the system's progress and growth over time. System progress and maturity models provide a visual representation of how fast or slow the system is growing. These models should be based on a five-year planning cycle with progress checks at the midpoint or two and a half years into the systems lifecycle.

The City of Oberlin, Ohio, has proven that there are no fixed rules or order when designing an enterprise project. Oberlin's IT manager felt that the datasets that she inherited were so disorganized that she decided to have two goal sessions. Basically, the first set of goals was to get the existing data organized, reviewed, and determine what was missing from the infrastructure information. When this first set of goals of accomplished then the city staff would do a normal cycle of goal selection.

There are times when only a couple of goals are needed to get a project moving forward. The author has witnessed a project where the goals were 1. Establish desktop GIS, 2. Produce a septic map, and 3. CAD to GIS. Each one of these goals by itself is easy to complete. When these three items are combined, they create the foundation for a new enterprise GIS. When the first set of goals have been accomplished then new goals should be established. A fellow GIS professional used an interesting cycle of selecting two to three goals as stepping stones toward establishing an enterprise GIS with the new company. This cycle of completing and selecting a small group of goals was an effective method of staying on course with the enterprise GIS build.

It is imperative that an assessment of the company resources and IT infrastructure is completed. This will provide a picture of the current state of the company. Determining what the company shortcomings are in hardware, software, skillsets, or even personnel will have an impact on the goals and determine a future state vision. Goals should be adjusted to where the primary goals are to address any of the company's present-day shortcomings that prevent it from supporting or enjoying the full benefit of an enterprise

GIS. This will also impact the future state since the future state will only be valid for the time required to hire new staff, acquire hardware, and software. When the company has fulfilled all the short-term goals, a new set of goals will be selected and adjustments reflecting these new goals will be added to the future state model. The heart and soul of any enterprise GIS or enterprise system is data being converted into information.

The only effective method to protect the data's integrity is the use of data standards. Data standards are not only essential for standardizing data collection, formats, procedures, and applications but crucial to establishing geodatabase domains. Geodatabase domains are necessary as the first line of defense against errors entering the database through editing, capture, and creation. Domains are essential tools for creating an accurate database and error-free dataset for the GIS.

The success of enterprise GIS depends upon clear, concise communications between staff, departments, and clients. Effective communication is a result of everyone understanding their positions and responsibilities in the enterprise GIS. These positions or roles establish a hierarchy of authority and accountability for the enterprise design team and company staff. The very top of the hierarchy is the GIS manager/enterprise architect with the GIS assistant manager/business architect assuming the role of an executive officer. The rest of the hierarchy answers to these two positions.

No enterprise GIS or EA system will succeed without support from the company's employees, management, or ownership. Convincing ownership and management to visibly support the development of the enterprise project will help achieve support from the rank-and-file employees. Any resistance about sharing data from certain employees can be overcome by offering them positions of data stewards and making them responsible for ensuring their data is updated and properly formatted. Listen and respond to staff concerns and questions, and invest the time in forming a good relationship with the organization's staff.

10.2 Leadership

Until now, this book has refrained from addressing the question, "Who is the person responsible?"

The answer to this rhetorical question is you, the reader. Sometime in the future of your career as a professional in information, GIS, data science, etc. there will come an opportunity to lead an enterprise type of project. Many professionals are nervous or afraid of becoming the point person on an enterprise project. The secret to being successful is quite simply be honest, friendly, and professional at all times.

A professional appearance, attitude, and knowledge will quickly establish a tentative trusting relationship with the client. The way to move from a tentative to a firm trusting relationship is by living up to one's word or better known as keeping a promise. Every project begins with a clean slate of trust, but missed deadlines, excuses, program shortcomings, misinformation, will chip away at this clean slate. Eventually, the broken promises result in the client terminating the consultant's contract or the project and finding a new consultant.

The consultant who invests time and effort into relationship building will enjoy a ROI that consists of many intangibles like employee insights, frustrations, and existing technical problems. Never overlook or dismiss staff opinions; these represent a treasure trove of valuable information for the system design team. Learning from the staff about past failures will prevent the consultant's team from committing similar mistakes.

Mistakes happen – that is a basic fact of life. A good leader can prevent a simple mistake from becoming a larger problem by projecting the impression of calm. Good leaders never let on that mistake has rattled them or made them nervous. Consider Admiral Adama from Battlestar Galactica. Adama suffered setbacks, made mistakes, and occasionally disciplined members of his crew. But he never let his crew or the general public witness him getting upset or losing control. The admiral did that only in private where no one could witness his self-doubt.

Do not play a blame game in front of the client or your team when confronted with a mistake. Calmly work through the problem and find a way to correct the mistake. Never publicly humiliate anyone for a mistake; if the mistake was preventable and a repeat offense, then take the person aside and have a private conversation, especially if you are at the client's office.

People will notice how a supervisor or manager treats his or her staff. When a leader is calm, controlled, and respectful to someone who made a bad mistake, people notice. A skilled consultant will use these fine points of leadership to gain his or her client's trust and respect, and, more importantly, to establish his or her credibility as a serious professional.

Proper planning and organization will produce an enterprise GIS that is adaptable, responsive, and efficient. When these two items are not the foundation of a GIS, the resulting system is a picture of chaos and confusion. A system in chaos cannot adapt, respond, or produce reliable analysis results; it creates mistrust and rejection among staff. This situation is preventable by using enterprise architecture principles to design and implement an enterprise GIS.

10.3 Future directions

Where will enterprise GIS or GIS be in the future? That is a hard question to answer. Every day brings a discovery, application, or new solutions to old problems. Today's digital world is globally connected and producing consumable information literally by the minute. The result is an explosion of an incredibly large number of enormous datasets that a new term was coined "Big data" to describe this phenomenon. Big data was begging to be mined, analyzed, and put to use.

GIS professionals jumped right in and began applying GIS capabilities to the vast pool of big data. Analyzing all of this information has allowed companies like Maxar (formerly DigitalGlobe) to support global initiatives to combat global poverty, child trafficking, and disaster response (MAXAR 2019). Universities have recognized the value of big data for GIS applications and research, Penn State has formed classes for geospatial big data analytics (Penn state web). The expressed purpose of these classes is to have GIS use big data to advance geographic research (PennState World Campus 2019).

There is a lot of potential in local government for the growth of enterprise GIS or GISs. Townships and small municipalities have smaller budgets and must find ways to get more value from today's dollar. Although enterprise systems require a little more investment on the front end for setup, down the road they offer a significant ROI for saving time and money. Demonstrating how enterprise GISs can save a municipality or township money, will normally get the consultant invited to make a presentation.

Public health's reliance on enterprise GISs and technologies will only grow over the next several years. Health departments at all levels – national, county, and city are trying to use data analysis and mapping to safeguard the public from disease. Hospitals are now experimenting with GIS to map the brain, analyze patient information, and keep track of patient locations within their buildings.

GIS-based online has stirred the public health interest in forming regional planning and response groups. Three or more health departments might join and share their information in the digital clouds and form regional public health planning groups. This type of data sharing and cooperation could lead to faster response times dealing with emergencies from storms, tornados, disease outbreaks, etc. A regional public health planning group supported by enterprise GIS could save a lot of lives.

The growth of online capabilities has led to a staggering amount of information sharing between people, countries, states, and communities. Dave Peters refers to this as the evolution of community GIS (Peters 2012). Community GIS uses the fantastic resources of the internet for cooperative projects, information sharing, and foster a global understanding of our fellow humans. GISCorps is a perfect example of a worldwide community.

Urban and Regional Information Systems Association (URISA) created the GISCorps to provide short-term GIS services to underprivileged communities (GISCORPS URISA 2019). The corps also provides emergency support to disaster response and relief efforts around the globe. Each member of the GISCorp is an unpaid volunteer who donates their mapping, programming, data analysis skills to help people in need. GISCorps website, https://www.

giscorps.org/, has a map that shows these people reside all over the world but use the internet to connect and work together to improve this world.

10.4 Final thoughts

This book's research and methodology should be considered a starting point for anyone desiring to enhance the value of enterprise GIS by working with enterprise architecture. This book has presented only the tools and areas of enterprise architecture that the author has used to aid municipalities in establishing organized and efficient enterprise GISs. There are many applications for enterprise architecture within GIS that is outside of this book's scope.

Enterprise GIS will impact and in turn be affected by these parts of enterprise architecture: data, technology, processes, events, etc. Research that models the event triggers and results in an enterprise GIS that would provide an almost immediate increase in efficiency to the information systems.

The author stated in the preface that this is not a book or a manual about enterprise architecture or geographic information systems. The purpose of this book is to provide a look at the methodologies and tools that were successful in enterprise system builds for the cities of Avon, Strongsville, Oberlin, Lyndhurst, and Brunswick, Ohio. Hopefully, readers have gained an understanding of where and how to implement enterprise systems for future municipalities. This author is attempting to honor his professors' teachings by "smoothing the path" for the professionals following him.

Bibliography

GISCORPS URISA. 2019. *Organizing Principles and Policies*. March 15. https://www.giscorps.org/what-we-do/organizing-principles-and-policies/.

MAXAR. 2019. *Global Development - Solutions for a Changing Planet*. March 5. https://www.digitalglobe.com/markets/global-development.

PennState World Campus. 2019. *Mater of Geographic Information Systems*. March 10. https://www.worldcampus.psu.edu/degrees-and-certificates/geographic-information-systems-gis-masters/overview.

Peters, Dave. 2012. *Building a GIS: System Architecture Design Strategies for Managers*. Redlands, CA: ESRI Press.

Appendix

Geographic Information System (GIS)
Chagrin Valley Engineering, Ltd.

Municipal Geographic Information System Design

Ryan Cummins, P.E. – Project Manager
John Woodard, M.S., GISP – GIS Specialist

Client:
City of Avon, Ohio

Client Contact:
Bryan Jensen, Mayor
36080 Chester Road
Avon, Ohio 44011

Approximate Project Budget:
Varies Annually

Start Date:
2015

Completion Date:
On-Going

As one of the most rapidly growing municipalities in northeastern Ohio, the City of Avon has recognized the need to utilize all available technologies to allow their staff to manage commercial, industrial and residential development in the City. One of the tools utilized by the City is GIS or Geographic Information Systems.

As the City's Consulting Engineer CVE, Chagrin Valley, Ltd. (CVE) is responsible for developing GIS for Avon. Our GIS staff has inventoried and created a folder-architecture for over 80 gigabytes of documents and plans for the City and is currently geo-referencing and digitizing the project.

CVE is using a balanced approach to assembling and implementing a GIS program over time that respects the municipal budget but quickly starts to produce operational and cost benefits. The system will be available to all City personnel and will include roadways, addresses, parcel data, sanitary sewers, water mains, storm sewers, zoning, aerial photographs, riparian zones, topography, FEMA flood zones, drug free school zones, streams, lakes, stormwater management facilities, commercial site plans, stormwater utility billing data, municipal parks & facilities, snow plowing routes, cemeteries, master bikeway plan, watershed development plan, as-built drawings and storm sewer outfalls. After receiving training from the CVE GIS staff, City personnel will be able to utilize the GIS systems to review infrastructure, zoning, recreation areas, development, etc. without having to leave their desktop and look for paper records.

When completed the Avon GIS system will provide tools that facilitate the acquisition, storage and management of data pools in relationship with geographic mapping. The resulting map products will enable complex data to be understood and disseminated easily. As such, Avon will be able to make vast amount of data available to the public upon request. The system will also provide a way to preserve institutional knowledge as municipal administrations and staff change over time.

This project was awarded "Best Practice 2016" by the state of Ohio Geographically Referenced Information Program OGRIP for an outstanding GIS program.

Geographic Information System (GIS)
Chagrin Valley Engineering, Ltd.

Municipal Geographic Information System Design

Jeff Filarski, P.E. – Project Manager
John Woodard, M.S., GISP – GIS Specialist

Client:
City of Lyndhurst, Ohio

Client Contact:
Rick Glady
Service Director
5301 Mayfield Road
Lyndhurst, Ohio 44124

Location:
Lyndhurst, Ohio

Approximate Project Budget:
Varies Annually

Start Date:
April 2006

Completion Date:
On-Going

In 2006, This project was undertaken to establish a consistent and coherent Geographic Information System (GIS) for the City of Lyndhurst's Building and Service Departments. The Building Department ensures that all buildings or structures meet local and state building codes. The Ser vice Department maintains city's infrastructure, while providing garbage and recycling services for city residents. Both departments recognized that a GIS system could be an important tool to aid city employees performing inspections or maintaining city assets.

Chagrin Valley Engineering, Ltd. (CVE) was contacted by the City of Lyndhurst and authorized to explore GIS solutions for the city departments. CVE recommended and them implemented a GIS system hosted and maintained on the company server. CVE staff then created a GIS viewer which is distributed to the city departments, that is user friendly and adaptable to meet the varied needs of each department. This viewer allows city staff to look up information for property owners by name, address, or parcel number.

Appendix

Enterprise Geographic Information System (EGIS)
Davey Resource Group, Inc.

Municipal EGIS system design

Client:
City of Oberlin, Ohio

Client Contact:
Ms. Dawn Ferro
I.S. Manager
85 South Main Street
Oberlin, Ohio 44074

Approximate Project Budget:

Phase 1 - 2019
$48,000.00

Phase 2 - 2020
$48,000.00

Start Date:
Phase 1
May 2019

Phase 2
January 2020

Completion Date:

Phase 1
December 2019

Phase 2
December 2020

John R. Woodard M.S., GISP. EA – GIS/IT Project Developer
William Ayersman M.S., GISP – Project Manager

The City of Oberlin, Ohio, recognized the need to utilize all available technologies to allow its staff to manage commercial, industrial, academic, utility, and residential development in the City. One of the tools used by the City is GIS or Geographic Information Systems.

As the City's Consulting GIS team, Davey Resource Group, Inc. "DRG" is responsible for developing an enterprise GIS (EGIS) for Oberlin. The GIS/IT staff has inventoried and created a folder-architecture for over 30 years' worth legacy data: documents, CAD drawings, plans for building the City's infrastructure.

DRG GIS/IT Team has a balanced approach to designing and implementing an EGIS program. This balanced approach over time respects the municipal budget but quickly starts to produce operational and cost benefits. The system will be available to all City personnel and will include roadways, addresses, parcel data, sanitary sewers, water mains, storm sewers, aerial photographs, streams, lakes, stormwater management facilities, storm sewer outfalls. After receiving training from the DRG GIS/IT staff, City personnel will be a ble to utilize the EGIS systems to review infrastructure, zoning, recreation areas, development, etc. without having to leave their desktop and look for paper records.

When completed The City of Oberlin's EGIS system will provide tools that facilitate the acquisition, storage, and management of data pools in relationship with geographic mapping. The resulting map products will enable complex data to be understood and disseminated quickly. As such, Oberlin will be able to make a vast amount of data available to the public upon request. The system will also provide a way to preserve institutional knowledge as municipal administrations and staff change over time.

This project was awarded "Best Practice 2019" by the state of Ohio Geographically Referenced Information Program OGRIP for an outstanding GIS program.

Geographic Information System (GIS)
Chagrin Valley Engineering, Ltd.

City of Strongsville, Ohio Enterprise Geographic Information System Design & Training

Client:
City of Strongsville

Client Contact:
Ken Mikula, P.E.
City Engineer City of Strongsville 16099
Foltz Parkway
Strongsville, Ohio
44149

Location:
Cuyahoga County, Ohio

Approximate Project Budget: $109,000

Start Date:
April, 2013

Completion Date: December 2013

Jeff Filarski, P.E. – Project Manager
John Woodard, M.S. – GIS Specialist

This project was undertaken to establish a GIS System for the City's Engineering Department in the City of Strongsville. The Engineer is responsible for all of the City's infrastructure planning. When the EPA mandated that Strongsville provide an inventory of outfalls and storm sewers within the City borders, it was decided to establish a GIS system. The City Engineer recognized the power and efficiency that a GIS system could provide to the organization and issued an RFP for a company that would be able to design a system that would be adequate for their varied needs, user friendly and scalable for future growth.

CVE designed, created, and oversaw the implementation of a storm, sanitary sewer and water distribution system GIS system based on the software application ArcGIS 10.2 from ESRI. This system is based on information digitized from geo-referenced plan sheets and is a complete inventory of the City's storm, sanitary and water systems. CVE's GIS Specialist worked closely with the City Engineer's staff both on site and remotely to design a system that supported current projects, satisfy EPA mandates, and become an information resource used by all City Departments.

The CVE GIS Specialist facilitated strategic planning meetings between the Engineer and IT/Data departments to ensure that a proper strategic vision was established for the Sewer and Water GIS System. This project then transitioned into an Enterprise GIS System that will be used by all City Departments. CVE's GIS Specialist then provided instruction and mentoring, offering advice and problem solving as the Engineer's staff increased their own GIS capabilities.

The final GIS system included a total of 205 miles of sanitary sewers, 227 miles of storm sewers, and 183 miles of water mains. In addition, 2,862 fire hydrants, 9,322 manholes, 5,014 catch basins, 1,567 yard drains, 848 headwalls, and 657 stormwater outfalls were digitally captured and placed in the system. All public and property information was included in the delivered GIS system.

CVE's approach resulted in a new GIS system that had the flexibility desired by the City Engineer and will be used by all City Departments, while seamlessly integrating into the existing IT/Data architecture. CVE continues to provide support services to The City of Strongsville.

Index

A

Accountability, 45, 127
Administrative assistants, 5
Application catalog, 101–104
Architecture/governance policy, 129–130
Asset Manager program, 127
Associated Business Capability, 102

B

BA, *see* Business architect
Backcasting, 83–85, 87, 89
Big data, 142
BOAP, *see* Business on a Page
Boy Scouts of America (BSA), 22
Brainstorming, 77, 78, 81
BSA, *see* Boy Scouts of America
Business architect (BA), 52–53, 129, 131, 136
Business capability, 33, 38, 77, 138–139
Business on a Page (BOAP)
    brainstorming, 77, 78, 81
    business architecture, 78
    development of, 78
    enabling level, 78
    operational level, 78, 80
    programming development, 80
    strategic level, 78

C

Card-based data models, 70
Catalogs, 96
Centralized database, 7
Chagrin Valley Engineering (CVE), 20, 21, 27, 85
Coherent system, 14
Communication hierarchy, 6
Community GIS, 142
Community Health Division, 34

Compliance waivers
    accuracy level, 134
    conformity, 132
    digitizing efforts, 135
    requirements, 133
    sample response, 133–134
Convention planning, 83
CVE, *see* Chagrin Valley Engineering

D

Data architect system model, 61
Data architecture, 3, 4, 53, 61–62
Data catalogs, 96–98
Data connections model, 119, 121
Data corruption, 63
Data errors, 28, 71
Data formatting, 73–74
Data management, 89
Data model, 8
Datasets, 73
Data sharing, 142
Data standards
    business intelligence analysis, 67
    data errors, 71
    definitions, 67
    field label format, 69
    geodatabase, 72
    information quality, 70
    legacy datasets, 71
    memory upgrade, 66
    object sensitivity, 69
    organizations, 67
    projection, 69
    reference cards, 67, 68, 70
    rules, 66
    tags (predefined), 68, 69
Data steward, 53–54
Data *vs.* information, 62
Data warehouses, 72
Davey Resource Group (DRG), 41, 126
Department supervisors, 57
Digital utility map, 85

Distributed GIS, 3
Document inventory, 86
Domain values, 65
DRG, *see* Davey Resource Group

E

EA, *see* Enterprise architecture
EA/EGIS tools
   backcasting, 83–85, 87
   BOAP, 77–81
   strategic analysis methods, 82–83
   UML, 81–82
   workflow, 81
EA roles and responsibilities
   accountability, 45
   architecture planning, 46
   combined GIS, 48, 49
   data models, 46
   enterprise system design process, 48
   executive level, 45
   executive management, 45, 48
   functional GIS department, 46, 48
   organizational chart, 49
   team building, 50–51
Enterprise architecture (EA), 4
   buzz word strategy, 10
   capabilities, 33–35
   communication, 11
   company organization, 11
   counter-productive situation, 11
   definition, 9, 10, 31
   deliverables, 11
   designing, 37–38
   EGIS, 1
   frustration, 32
   geographically distributed network, 10
   GIS, 2
   goals, 38–42
   integration, 10
   issues, 32
   organization's business systems, 2
   roles and responsibilities (*see* EA roles and responsibilities)
   valuable information, 32
Enterprise data, 62–63
Enterprise Data Model, 63

Enterprise geographic information system (enterprise GIS/EGIS)
   assessment, 26, 27
   capabilities, 17
   centralized database, 17
   data errors, 28
   datasets, 28
   description, 1
   EA tools, 91
   elements, 25
   evaluation, 26
   executive management, 27
   geodatabase, 92–95
   goals, 91–93, 110
   implementation, 25
   maturity table, 92, 94
   old-timer knowledge, 28
   organizational support, 18
   organizations, 12–13
   ownership, 27
   potential clients/opportunities, 26
   software vendors, 18–19
   transitions, 18, 25
Enterprise GIS for Local Government, 8, 9
Enterprise-level GIS transportation systems, 8, 28
Enterprise System Design Planning Tools, 7
Environmental Systems Research Institute (ESRI), 6, 18
   ArcGIS software, 29
Executive secretaries, 56

F

FGDC National Data Standards and Publications, 67
FGDC National Standards Working Group, 67
Financial analysis, 5
Flexible architecture, 35–36
Future vision, 105

G

Geodatabase, 92–95, 137, 140
   architecture model, 120

Index

classes, 72
domains, 73–75, 95
feature dataset, 72
Geographic information systems
    (GIS), 78
  application/technology architect,
    55–56
  capabilities, 12
  distributed, 3
  EA, 2
  factors, 8
  professionals, 2
Geoidentifier, 77
Geo-referencing, 86
Geospatial Consortium (OGC), 17
GIS, see Geographic information
    systems
GISCorps, 142
GIS manager/enterprise architect
    (GISM-EA), 51–52
GISM-EA, see GIS manager/enterprise
    architect
Goal assessment, 139
Governance
  assignment, 136
  business units, 126
  corruption, 125
  data architecture, 126
  principles, 127
  requirements, 125
Governance review board, 130–132

I

Impartiality, 128
Improvising tools, 88–89
Information management, 138
Information sources
  encyclopedia, 7
  ESRI, 6
  GIS capabilities, 6
  resource pages, 7
IT infrastructure, 139

L

Leadership, 140–141
Legacy data, 95
Look-up table, 72–73, 95

M

Margin, 82, 83
Master data management (MDM), 71
Maturity models, 107–111, 139
MCHD, see Medina County Health
    Department
MDM, see Master data management
Medina County Health Department
    (MCHD), 19–21, 34, 128
Metadata, 64
Miscommunication, 28

N

NASCIO, see National Association
    of State Chief Information
    Officers
National Association of State Chief
    Information Officers
    (NASCIO), 77, 89, 109
Navy Mobile Construction Battalion
    (NMCB), 22
NMCB, see Navy Mobile Construction
    Battalion

O

Oberlin Municipal Light and Power
    Service (OMLPS), 126, 127, 132
Object Management Group (OMG), 81
Object sensitivity, 69
Ohio Turnpicke Commission (OTC),
    116, 117
OMG, see Object Management Group
OMLPS, see Oberlin Municipal Light
    and Power Service
Online GIS, 128
Online resources, 12, 88
OpenSource Enterprise GIS, 7
Open Source Geospatial Foundation, 7
OTC, see Ohio Turnpicke Commission
Owner and executive management, 27,
    56–57

P

Parcel map, 4
Personalized formatting, 73

Person recruiting exhibitors (vendors), 84
Primary activities, 82
Process model, 88
Professionalism
 client's trust, 23–24
 contractor's assumption, 21
 dismissal, 22
 honesty, 24
 knowledgeable professional, 22
 system design, 23
Professional surveyor, 4
Property boundaries (parcel lines), 3
Proprietary software, 2
Public Health and Emergency
 Management, 13
Public health interest, 142
Public health officials, 71
Public health's reliance, 142
Public Works Department, 126

R

Reference data, 65, 72–73
Return on investment (ROI), 2, 32, 98, 138, 141
ROI, *see* Return on investment

S

Seamless integration, 29
Septic system, 5
SMEs, *see* Subject matter experts
Software packages, 138
Specialization, 83
Star Trek Next Generation (STNG), 52, 62
STNG, *see* Star Trek Next Generation
Strategic architectures, 115–116, 119, 188
Strengths, weaknesses, opportunities and threats (SWOT), 77, 82–83
Sub-architectures, 116, 119, 122–124
Subject matter experts (SMEs), 49, 54–55
Summit County Health Department, 31
Support activities, 82
SWOT, *see* Strengths, weaknesses, opportunities and threats
System failures, 1
System governance, 13, 129
System hierarchies, 64–65
System planning, 138
System progress model, 139

System visualization
 diagramming programs, 118
 enterprise system, 115
 fiscal planning, 118
 geodatabase, 119
 internal/external connections, 115
 strategic architectures, 115–116, 118
 valuable resources, 115, 119

T

Table to domain tool, 95
Technology catalogs, 98–101
The Open Group Architecture
 Framework (TOGAF), 10
The Original Series (TOS), 51, 62
TOGAF, *see* The Open Group
 Architecture Framework
TOS, *see* The Original Series
Transparency, 127

U

UML, *see* Unified modeling language
UML activity diagram, 81–82
Unified modeling language (UML), 81
Universal data model, 29
Urban and Regional Information
 Systems Association
 (URISA), 142
URISA, *see* Urban and Regional
 Information Systems
 Association
USGS National Geospatial Program
 Standards and Specifications, 67
US Integrated Taxonomic System
 (ITIS), 67
US National Vegetation Classification
 (USNVC), 67
USNVC, *see* US National Vegetation
 Classification

V

Value chain analysis, 82

W

West Nile Virus (WNV), 34
Workflow model, 81